ECRR

Chernobyl: 20 Years On

Health Effects of the Chernobyl Accident
European Committee on Radiation Risk
Documents of the ECRR 2006 No1

Edited by
C.C.Busby and A.V.Yablokov

Published on behalf of the European Committee
on Radiation Risk
Comité Européen sur le Risque de l'Irradiation

Green Audit ❖ 2006

European Committee on Radiation Risk
Comité Européen sur le Risque de l'Irradiation

Secretariat: Greta Bengtsson, Grattan Healy
Scientific Secretary: Chris Busby
Website: www.euradcom.org
Contact: admin@euradcom.org

ECRR
Chernobyl 20 Years On; The Health Effects of the Chernobyl Accident
Documents of the ECRR 2006 No 1
Edited by:
Chris Busby, Alexey Yablokov

Published for the ECRR by:
Green Audit Press, Aberystwyth, SY23 1DZ, United Kingdom
Copies of this book can be ordered directly from
admin@euradcom.org or from any bookseller

First Edition: April 2006
ISBN: 1-897761-25-2
A catalogue for this book is available from the British Library
Printed in Wales by Cambrian Printers

(Cover: Boy with Thyroid Cancer in Belarus- Photo: Andrew Testa)

What do we really know today? Really not too much... Chernobyl produced 31 deaths and 2,000 avoidable thyroid cancers in children. No other internationally confirmed evidence exists.

Abel Gonzalez.
Representative of the International Atomic Energy Agency (IAEA)
Speaking at WHO conference on Chernobyl Effects in Kiev, Ukraine, 2001.

The risk of leukaemia doesn't appear to be elevated even among the recovery workers. No scientific evidence either for increases in cancer incidence or other non-malignant disorders that could be related to the accident exists.

Norman Gentner.
Representative of the United Nations Scientific Committee on the Effects of Atomic Radiation (UNSCEAR). Speaking at WHO conference on Chernobyl Effects in Kiev, Ukraine, 2001.

The consequences of Chernobyl do not fade away but actually grow increasingly uncertain and in many ways more intense. I agree with the UN Secretary General Kofi Annan, who said, 'the legacy of Chernobyl will be with us and our descendants for generations to come'.

D.Zupka
Representative of the United Nations Office for Humanitarian Affairs (OCHA). Speaking at WHO conference on Chernobyl Effects in Kiev, Ukraine, 2001.

They were deliberating among themselves as to how they could give Wings to Death so that it could, in a moment, penetrate everywhere, both near and far.

Jan Amos Komensky (Comenius)
The Labyrinth of the Worlds (1623)

'Run like a dog and flee like a hare'

This volume is dedicated to the Chernobyl 'liquidators'.
They gave their lives so that Europe could remain inhabitable.
They were poorly repaid.

Contents

Introduction

One of the first sub-committees formed by the European Committee on Radiation Risk (ECRR) was the Chernobyl sub-Committee. Its remit was to examine the epidemiological and other evidence on the health effects of low dose radiation exposure which could be obtained from careful study of populations living in areas contaminated by the Chernobyl accident. The ECRR had been founded on the basis that many scientists had criticised the external acute exposure foundations of the current radiation risk model (employed by all countries for radiation protection purposes), the model, essentially, of the International Commission on Radiological Protection (ICRP). This model, it was argued, was scientifically invalid for internal chronic irradiation from fission product isotopes and radioactive microscopic particles.

It seemed to the committee that the Chernobyl catastrophe represented a unique opportunity or natural laboratory to examine the health effects of such low-dose internal exposures. Knowledge obtained from such studies would be valuable in developing an accurate understanding of the effects of radiation, and also in interpreting the many reports of apparent associations between cancer, leukaemia or other ill health and prior exposures to internal radionuclides e.g. from weapons test fallout, and nuclear site discharges. Indeed, the question of the adequacy of the ICRP external risk model had been called into question for many years by a number of independent scientists and, with the rapid developments of radiation biology in the 1990s following the discovery of 'genomic instability', by the early 2000s calls were increasingly being made to re-examine the issue. In 2001, following much anecdotal evidence of increases in ill health in the Chernobyl affected territories of the ex-Soviet Union, and also reports of increases in infant leukaemia in European countries in the children who were in the womb over the period of the internal exposures, the European Parliament called for the re-examination of these models specifically in connection with the effects of the Chernobyl accident.

Also in 2001, in the UK, a new committee was set up under the joint direction of the Department of the Environment and the Department of Health. The remit of this Committee Examining Radiation Risk from Internal Emitters (CERRIE) was to examine just these issues. In 2003 CERRIE organised an international workshop in Oxford and most of the major radiation risk experts in the world were invited to attend to discuss the issues and comment on the draft reports. Most did so. Among those who came were the eminent scientists Professors Alexey Yablokov, Elena Burlakova and Inge Schmitz Feuerhake. The Russian scientists drew attention to the many studies reported in the Russian language literature. These research papers on the Chernobyl effects were not being translated into English by the UN agencies, nor by the World Health Organisation (WHO). For this reason (they said) the terrible effects of low dose radiation in the Chernobyl affected territories were being ignored or glossed over. Surely CERRIE should attempt to examine the truly enormous amount of useful information that these Russian language reports represented? In the event, the CERRIE secretariat did nothing and the CERRIE committee ended in 2004 split on the issue of internal radiation with two reports being published in late 2004. Only the Minority CERRIE report reviewed some fifty of the main Russian studies and drew attention to the serious cancer and non-cancer effects following Chernobyl reported in the Russian journals.

It is now the 20th anniversary of the accident and in the West nothing has changed. It as if none of these events ever occurred. Children continue to die of cancer near nuclear polluted sites, which still continue to release fission-product radioisotopes under licenses based on the IRCP model. Court cases are still lost by the enraged and desperate parents because judges still believe that the doses to the children were too low. Scientists on government committees still talk about 'absorbed dose' as if it were a meaningful concept for internal irradiation. The Emperor still wears his new clothes.

The evidence from the Chernobyl affected territories, presented here in these chapters, reveals the real-world consequences of a simple and terrible new discovery: that the effects of low dose internal irradiation cause subtle changes in the genome that result in an increase in the general mutation rate. This genomic instability was first seen in cells in the laboratory. The Chernobyl evidence, presented here, shows that this seems to be true for all species, for plants and animals and humans. It has profound implications that go beyond radiation protection and risk models. In the review paper by Krysanov in this collection we find that mice living in the high irradiation zone, 22 generations after the initial exposure, are *more* radiosensitive than mice living in lower exposure areas. The same effect is reported for plants by Grodzhinsky who wryly points out that plants cannot exhibit the 'radiophobia' that many of the Chernobyl effects have been blamed on. This flies in the face of current ideas about genetic selection.

The effects of genomic instability are apparent in the evidence of massive harm to the organs and systems of living creatures at low doses of internal exposure, resulting in a kind of radiation ageing associated with random mutations in all cells. At the higher doses in the 'liquidators', after some years, their bodies seem to simply fall apart. In an astonishing statement we hear from Yablokov that in Moscow 100% of the liquidators are sick, in Leningrad 85%. These are men that ran like hares into the radiation fields with improvised lead waistcoats cut from roofs and who, by stabilising the situation at the reactor, saved Europe from a nuclear explosion equivalent to 50 Hiroshima Bombs - an outcome that would have made most of it uninhabitable. They are forgotten.

Whole biological systems collapse; at the cell level, at the tissue level and at the population level. Burlakova and Nazarov describe these subtle effects at lower doses of internal irradiation in laboratory cell systems and also people, Grodzhinsky shows the effects in plants, - higher for internal exposures than external, Krysanov shows the effects in wild animals and Yablokov and the Nesterenkos in the children and adults living and continuing to live in the contaminated territories. The effects clearly operate at what are presently thought to be vanishingly low doses. The increases in infant leukaemia in several countries in Europe flag up the extreme dissonance between the IRCP model and the true effects. This finding has been ignored by the WHO. The papers are not referenced in UNSCEAR or in the recent US BEIR VII report. The comparison between the expected and observed cases of infant leukaemia gives a clear indication of an error of upwards of 150-fold in the current model's prediction. This is shocking. It means that the previous releases to the environment from accidents, from weapons tests or under licence have killed and will still kill millions. The effects of the 1960s atmospheric weapons tests are with us and our children forever and are clearly responsible for many of the current illnesses.

It is a scandal that UN agencies charged with protecting the public - e.g. the United Nations Scientific Committee on the Effects of Atomic Radiation (UNSCEAR) or the World Health Organisation (WHO) - can ignore the huge amount of evidence from the Chernobyl accident that shows these effects. This evidence has been presented to them again and again. Yet in the WHO conference on Chernobyl in Kiev in 2001 the representative of UNSCEAR, Norman Gentner stated clearly:

The risk of leukaemia doesn't appear to be elevated even among the recovery workers. No scientific evidence for increases in cancer incidence or other non-malignant disorders that could be related to the accident.

Science moves forward through experiment and through observation. The radiation risk model presently used to underpin legal constraints on public exposure is based mainly on the theory that the external acute radiation effects in the Japanese A-Bomb survivors can be used to predict and explain effects from exposure to internal novel fission-products that never existed on earth for the whole of evolution. The Chernobyl accident and its appalling outcomes have given the human race the empirical evidence to test this theory. The observations made or reviewed in these extraordinary chapters - many written by eminent scientists-makes it fundamentally clear that the present radiation risk model is flawed.

The ECRR sub-committee on Chernobyl has worked hard under difficult conditions to assemble the contributions from these eminent scientists and to put them into reasonable English. This book represents a landmark on the road to understanding the effects of low-dose chronic irradiation. The committee believe that these lessons should be borne in mind by policy makers who are, even now, discussing new investments in nuclear energy and ways in which historic and future radionuclide waste can be disposed of into the environment. The committee recommends this book to scientists and policymakers and concerned members of the public in the hope that the huge amount of work carried out by scientists publishing their results in Russian language journals and others studying the effects of the Chernobyl accident will influence their decisions in this important area of public health.

The Committee thanks Greta Bengtsson, Mireille de Messieres and Saoirse Morgan for their hard work in the preparation of this book.

Chris Busby, Scientific Secretary, ECRR

CHAPTER 1

The Chernobyl Catastrophe - 20 Years After (a meta-review)

Alexey V. Yablokov

Russian Academy of Sciences, and Center for Russian Ecological Policy, Moscow

The first official forecasts regarding the consequences of the explosion of the reactor in the 4th block of the Chernobyl NPP on April 26th, 1986 concerning the health of the population of the USSR predicted only several additional cases of cancer over some tens of years. In 20 years it has become clear that not tens, hundreds or thousands, but millions of people in the Northern hemisphere have suffered and will suffer from the Chernobyl catastrophe. These people include (Grodzinsky, 1999 *et al*):

1. More than 220,000 residents of highly contaminated areas who were evacuated in 1987 across Belarus, Russia and Ukraine.
2. Those who received significant doses of irradiation during the first days and weeks, and/or living in territories with radioactive pollution more than 1 Ci/km2 (in Ukraine - up to 3,2 millions, in Russia - up to 2,4 million, in Belarus - up to 2,6 million persons, in Sweden, Norway, Bulgaria. Romania, Austria, Southern Germany and a number of other countries across Europe - 0,5 - 0,8 million persons).
3. Liquidators (the persons who took part in operations to minimize the consequences of the catastrophe both at the NPP and in the polluted territories nearby): 740 thousand persons from Ukraine, Russia and Belarus, and 80 - 90 thousand from Moldova, the Baltic's states, the Caucasus, Middle and Central Asia;
4. Children, whose parents belong to the first three groups: Up to 2,006 approx. 1 - 2 million.
5. People living in territories where the Chernobyl fallout fell, basically, in the Northern hemisphere (including Europe, North America, Asia): A number difficult to define, but not less than 2,500 million persons;
6. People who have consumed the Chernobyl fallouts' radioactive-polluted foodstuffs (basically, in the countries of the former USSR, but also in Sweden, Norway, Scotland and a number of other European countries) – possibly in the order of several hundred thousands.

For the first of these four groups exposed to the consequence of the Chernobyl disaster additional irradiation is (or can be) substantially determined, for the fifth and sixth groups - stochastic.

Official secrecy (until May 23rd, 1989) and irreversible state falsification of medical data during first three years after the catastrophe, as well as an absence of authentic medical statistics in the former USSR, highlights the inadequacy of material concerning primary epidemiological consequences of this catastrophe. There is also uncertainty in determining the quantity of radionuclides discharged from the reactor: from 50 million Ci (Soviet official data) up to 3,500 millions Ci (several independent estimations).

There are some principal difficulties in establishing a direct connection between levels of irradiation and health effects. These difficulties include:

*   extremely localized concentrations of fallout:
*   difficulty establishing the dose of radiation exposure caused by short-lived radionuclides during the first hours, days and weeks after catastrophe (I-133, I-135, Te-132 and a number of others);

- the behaviour of "hot particles»;
- too complex a picture of the radionuclides' transformations, migration and bio-concentrations in ecosystems;
- little known specific effect of each radionuclide (e.g. pollution by Sr-90 has consequences for the immune system, pollution by Cs-137 - others under identical density of radiation, Evetz et al., 1993);
- different biological effects of internal and external irradiation (e.g. internal irradiation leads to a gradual auto-immune reaction, whereas external irradiation leads to a fast one. Lisianyi, Lubich, 2001).

The problems listed above make it difficult to reconstruct individual doses and dose rates, and cast doubt on any reports of a "strong correlation" between levels of the Chernobyl irradiation and specific health effects.

Further difficulties in understanding the real consequences of the Chernobyl radioactive fallout include numerous, scientifically unproven statements made by representatives of the nuclear industry, and experts connected with it, about the insignificance of the catastrophe for public health (e.g., "Chernobyl Forum Report", September 2005).

In order to reveal the full consequences of the Chernobyl catastrophe it is essential to discover its true influence on public health by comparing the same groups during the same periods after the catastrophe and comparing populations in territories, identical in geographical, social and economic conditions and differing only by the level of irradiation.

1. Mortality

Since 1986, in the USSR, life expectancy has noticeably decreased. On average, infant mortality has noticeably increased, as well as death rates for those of advanced ages. There is no proof of a direct connection between these parameters and the Chernobyl catastrophe, but there is proof of such connections for particular polluted territories.

After 1986, in the radioactively polluted areas of Ukraine, Belarus and Russia, there is an increase in general mortality by comparison with neighboring areas (Grodzinsky, 1999; Omelianetz, et al., 2001; Kashirina, 2005; Sergeeva et al., 2005).

An increase in the number of stillbirths, correlating with the level of pollution, is noted for of some areas of Ukraine (Kulakov et al., 1993) and Belarus (Golovko, Izhevsky, 1996). The number of spontaneous abortions (miscarriages) has noticeably increased in some polluted territories of Russia. By some estimates, the number of miscarriages and stillbirths as a result of radioactive pollution in Ukraine has reached 50,000 (Lipik, 2004).

In some European countries an increase in perinatal mortality connected with the Chernobyl catastrophe has been revealed (see Korblein paper in this book). After 1987 an increase in infant and children's mortality is noted in the polluted areas of Ukraine (Omelianetz, Klement'eva, 2001) and Russia (Utka et al., 2005). Presented in Table 1 is one example of such mortality in one Russian area.

Table 1 Infant (per 1,000 live births) and general mortality (per 1,000) in three radioactively polluted administrative districts of Bryansk area, Russia in 1995 - 1998 (the Condition ..., 1999; Komogortseva, 2001).

	Three polluted districts*				Bryansk area	Russia
	1995	1996	1997	1998	1998	1997
Infant mortality	17.2	17.6	17.7	20.0	10,2	17.2
General mortality	16.7	17.0	18.2	21,4	16.3	13.8

Novozybkovsky, Klintsovsky, Zlynkovsky.

Cancer mortality for 1986 - 1998 has increased by 18 - 22 % in radioactively polluted Ukrainian territories and among the evacuees, (compared with the whole of Ukraine - 12 %) (Omelianetz *et al.*, 2001; Golubchikov *et al.*, 2002). Mortality in men from prostate cancer has increased in the Ukrainian polluted territories by 1,5 - 2,2 times (in Ukraine as a whole - by 1,3 times) (Omelianetz *et al.*, 2001). It has been revealed that in the radioactively polluted territories of Belarus the majority of sudden deaths correlated with the level of radionuclide incorporation (Bandajevsky, 1999).

These facts that establish a general increase in the level of child and infant mortality in the radioactively polluted territories (and to a greater degree on the more polluted territories) when compared with similar data in uncontaminated territories, leave no doubt that absence of these data for the whole area of the Chernobyl fallout region has hitherto been associated with incorrect statistical data.

2. Cancers

Belarus showed a 40 % increase in cancer between 1990 and 2000. Thus, in the territories most radioactively polluted (the Gomel area) the increase was maximal, and in the least polluted (the Brest and Mogyliov areas) - minimal (Okeanov et al., 2004). In 1987 - 1999 in Belarus about 26,000 radiogenic cancers (including leucosis) were noted; from these cases 11,000 have died (Mal'kov, 2001). Average value of excess absolute cancer risk was 434/104 person/years/Sv (relative risk - 3 - 13 x 10^{-1}), which is above the limit accepted by UNSCEAR (Mal'kov, 2001).

Table 2 presents calculations of cancer rates based on a collective dose of all generations for the entire period of additional Chernobyl Cs-137 irradiation.

For the general number of people in different countries who will have cancers from the Chernobyl Cs-137 during their lifetime (about 951,000 persons, see Table 2), it is necessary to add the number of cancer cases as a result of irradiation by I-133, I-135 (mostly thyroid cancers) and more than 25 other short-lived nuclides, including strontium, plutonium and americium, uranium and hot particles.

Table 2 Calculated number of cancer cases (without leukaemia) for all generations *, caused by Chernobyl Cs-137 (Goffmann, 1994)

	Number of cases	
	Fatal	Non- fatal
CIS, European part	212 150	212 150
The Europe (without the CIS)	244 786	244 786
Other countries of the world	18 512	18 512
IN TOTAL	475 368	475 368

* *On the basis of an expected collective dose "indefinitely" in 127.4 million person/Rad.*

Although other estimates of the number of fatal cancer cases induced by the Chernobyl irradiation claim "only" 22,000 – 28,000 deaths, J. Goffmann (1994) convincingly reveals a clear understatement on behalf of the authors, or the unreasonable understating of the collective dose, based on underestimates of emissions from the blown up reactor. Table 3 presents some examples of the research that has shown connections between occurrence of some cancers and the Chernobyl pollution.

Before the Chernobyl catastrophe cancer of the thyroid gland in children and teenagers in territories in Ukraine, Belarus and Russia rarely occurred. In Belarus only 21 cases were registered between 1965-1985 (Demedchik *et al.*, 1994), in Ukraine, before the catastrophe, no more than 5 cases were registered annually, in nearby Russia - 100. In the year 2000 the number of cases of this cancer in children and teenagers in the polluted territories had increased by hundreds of times. 4,400 cases of radiation-induced thyroid cancer have been recorded in Belarus (Malko, 2002) and about 12,000 cases of thyroid cancer are considered to have already appeared in the affected three countries (Imanaka, 2002). This differs from the UNSCEAR 2000 estimation of about 1,800 thyroid cancers observed during 1990 - 1998 in children 0 - 17 years old in 1986, and differs from the Chernobyl Forum Report's (2005) estimation of about 4,000 cases.

In Belarus the relative risk of radio-induced thyroid cancer (ERR) has exceeded more than the risk factor of 8x 10^{-2} Gy^{-1} from ICRP-60 (Malko, 2004).

Based on the dynamics of radiogenic thyroid cancer growth and the character of the pollution, it is possible to assume that during the following 40 - 50 years in Ukraine there will be about 30,000 additional cancer cases, in Belarus – 50,000, and in Russia – 15,000. There are also reports of a rise in cases of thyroid cancer in the South of France, Scotland and Poland.

Because the latent period for leucosis is between several months to several years, many cases in Ukraine, Belarus and Russia have never been registered; this is also due to official orders to falsify such data. In spite of this, there is a visible increase in the frequency of leukemia in all polluted areas of Ukraine, Belarus and Russia (Prysyazhnyuk *et al.*, 1999; Ivanov *et al.*, 1996; Sources and Effects ..., 2000).

Table 3 Examples of occurrence of some solid cancers cases as a direct result of the Chernobyl catastrophe

Location	Region, features	Author
Retinoblastoma	A 2-fold increase in cases between 1987 and 1990 in the eye microsurgical center in Minsk, Belarus	Byrich *et al.*, 1994
Lung	A 4-fold increase among 32 000 evacuees, than on the Belarus' average	Marples, 1996
Intestines, Colon, Kidneys, Lungs, Mammary glands, Bladder	An increase in the Gomel area (Belarus), correlated with a level of the Chernobyl radioactive pollution	Okeanov, Yakimovich, 1999
Respiratory organs	An increase in the Kaluga area (Russia), correlated with the Chernobyl radioactive pollution	Ivanov *et al.*, 1997
Bladder	An increase in men in the Chernobyl polluted territories in Ukraine	Romanenko *et al.*, 1999
	Increase in the liquidators in Belarus	Okeanov *et al.*, 1996
Nervous system	Increased by 76,9 % from 1986 to 1989	Orlov *et al.*, 2001
All cancers	An increase (from 1,34 % in 1986 to 3,91 % in 1994) among adults from the polluted territories of the Zhytomir area, Ukraine	Nagornaya, 1995
Pancreas	Up to 10-fold increase in the most polluted areas Ukraine, Belarus and Russia from 1986 to 1994	Sources and effects … 2000
Mammary gland	Increase of 1,5 in polluted Ukrainian territories for period 1993 - 1997	Москаленко, 2003).
All cancers in children	In the 11 yr period after the catastrophe, for the most polluted areas, rates (13,1 - 17,1 per 100,000 were higher than the Russian average (10,5)	Ushakova, *et al.*, 2000).
	Exceeds the average across Belarus by 3,7 – 3,1 times for the evacuated children and those living in the polluted regions	Belookaya *et al.*, 2002
	Exceeded by 20 times in 1994 in the Gomel area (heavy polluted), than in the less polluted Vitebsk area, Belarus	Bogdanovich, 1997
	Exceeds up to 15 times in 1995-1996 compared with the period 1968 – 1987 in Lipetsk city, Russia	Krapyvin, 1997

3. Diseases of the Nervous system

Presented in Table 4 are data on the level of illnesses of the nervous system for irradiated Ukrainian territories. Table 5 presents research that has shown a correlation between levels of radioactive pollution and psychological diseases.

Table 4 Dynamics of diseases of the nervous system for the period 1987 - 1992 (per 100,000, adults) in territories within Ukraine, as a result of the Chernobyl catastrophe (Nyagu, 1995)

Years	1987	1988	1989	1990	1991	1992
Diseases of nervous system	2641	2423	3559	5634	15041	14021
Mental diseases	252	419	576	1157	5114	4931

Table 5 Examples of the occurrence and level of morbidity of psychological diseases in the Chernobyl polluted areas

Diseases	Area, feature	The author
Congenital convulsive syndrome	Growth over 10 years in radioactively polluted areas of Belarus	Tsymlyakova, Lavrent'ieva, 1996
Brain circulation pathology	Occurrence 6 times greater in a group of agricultural machine operators from heavily polluted districts of the Gomel area, Belarus (at 27,1 % from 340 against 4,5 % from 202 in the control group)	Ushakov et al., 1997
General neurological diseases, a volume of short-term memory loss, deterioration of attention function in school-children, 16 - 17 years.	Increase in the polluted districts of the Mogyliov area, Belarus	Lukomsky et al., 1993

There is more and more evidence of a "Chernobyl dementia" phenomena (deterioration of memory and motor skills, occurrence of convulsions, pulsing headaches), caused by the destruction of brain cells in adult people (Sokolovskaya, 1997).

4. Cataract

In the Chernobyl territories cataracts has become a common disease. In Belarus, especially, it often occurs at a level of pollution above 15 Ci/km2 (Paramey et al., 1993; Edwards, 1995; Goncharova, 2000, Table 6).

Table 6 Occurrence of cataracts at different levels of pollution in 1993 – 1994, Belarus (Goncharova, 2000)

Year	All Belarus	1-15 Ci/km2	More than 15 Ci/km2	Evacuees from a zone more than 40 Ci/km2
1993	136,2	189,6	225,8	354,9
1994	146,1	196,0	365,9	425,0

5. Urogenital system diseases

Table 7 presents examples of studies revealing a correlation between the level of radioactive pollution and the urogenital illnesses.

Table 7 Examples of diseases of the urogenital system in the Chernobyl radioactive fallout territories

Disease	Features, area	Author
Interruption of pregnancy, Gestosis, Premature birth	Increase in evacuees and those living in the polluted territories for 8 - 10 years	Golubchikov *et al.*, 2002; Kyra *et al.*, 2003
Inflammation of female genitals	Increased in Ukraine through 5 - 6 years after the catastrophe	Gorptchenko *et al.*, 1995
Ovary, cyst, uterus, fibroma	2-fold increase in Ukraine 5-6 years after the catastrophe	
Menstruation irregularities	3-fold increase compared with the period before the catastrophe (initially, strengthened menstruations prevailed, after 5 - 6 years - poor and rare).	Gorptchenko *et al.*, 1995
	In the majority of women of childbearing age of Belarus and Ukraine	Nesterenko *et al.*, 1993; Vovk, Mysurgyna, 1994; Babytch, Lypchanska, 1994
Kidney infections, stones in kidney and in urine passages	Increase and prevalence among teenagers, Ukraine	Karpenko *et al.*, 2003
Illnesses of urogenital system	Higher level among 1 026 046 mothers of newborns on territories with more than 1 Ci/km2, Belarus	Busuet *et al.*, 2002
Infringements of sexual development	Increased cases in the polluted territories, Belarus	Sharapov, 2001
	5 times higher level of infringements in girls, 3 times - in boys in the heavily polluted territories than on	Nesterenko *et al.*, 1993

	less polluted, Belarus	
Gynecologic disease, complications of pregnancy and births	Increased during 1991-2001 in Belarus	Belookaya *et al.*, 2002
Failures of pregnancy, medical abortions	Growing in the polluted areas of Belarus	Golovko, Izhevsky, 1996
Infertility	Increase of 5,5 times in the polluted Belarusian areas in 1991 in comparison with 1986	Shilko *et al.*, 1993.
Pathology of sperm	Increase of 6.6 times in the polluted areas, Belarus	Shilko *et al.*, 1993
Sclerocystosis	2-fold increase in the polluted areas, Belarus	Shilko *et al.*, 1993
Early impotence in men (aged 25 - 30 years)	Increase in the polluted areas of Belarus and Russia	Shilko *et al.*, 1993;
Structural changes of testiculus and spermatogenesis disturbancies	In 75,6 % of the surveyed men in Kaluga area, Russia	Pysarenko, 2003
Lactation in 70-year old women	Belarus	Alexievich, 1997
Delay of puberty	Delayed in young men (for two years) and girls (for a year) in the areas polluted by 90-Sr and plutonium;	Paramonova, Nedvetskaya, 1993
Acceleration of sexual development	Girls (13-14 years) in the territories polluted by 137-Cs	Paramonova, Nedvetskaya, 1993; Leonov, 2001

6. Cardio-vascular system and blood diseases

Diseases of the cardio-vascular system and blood are one of the most common consequences of the Chernobyl radioactive pollution (Table 8).

Table 8 Diseases of the cardiovascular system in Chernobyls' radioactively polluted areas

Disease	Region, features	The author
Anaemia	Increase of 7 times in the Mogyliov region; correlates with the level of pollution in Belarus	Hoffman 1994, p. 514; Dzykovitch et al., 1994; Nesterenko, 1996
Illnesses of the blood circulation system	Level of primary illnesses have risen by 2,5 - 3,5 times in Mogyliov and Gomel areas, Belarus since 1986	Nesterenko, 1996
	Increase of 3-6 times more in the polluted districts, than the average in the Bryansk area	Komogortseva, 2001
Arterial hypertensia or hypotensia	More often in territories with a level of pollution above 30 Ci/km2 in the Mogyliov area	Podpalov, 1994;
	In children, and adults correlated with the pollution level, Belarus	Nedvetskaya, Lyalykov, 1994; Sykorenskyi, Bagel', 1992; Goncharik, 1992; Zabolotny et al., 200
Disturbances of the heart' rhythm and the digestive system	Ischemic illness more frequent and severe in the polluted areas, Belarus	Arynchyna, Mil'kamanovitch, 1992.
	Correlated with the level of 137-Cs incorporated	Bandajevsky, 1997, 1999
	Increase in the polluted territories of Belarus	Nedvetskaya, Lyalykov, 1994; Sykorenskyi, Bagel', 1992; Goncharyk, 1992
Macrocitosis of lymphocytes	6-7 times more often in the polluted areas, Belarus	Bandajevsky, 1999
Diseases of the blood and circulatory organs in adults	Increased by 50 times in the polluted territories from 1986 to 1994 in the Zhytomir area, Ukraine	Nagornaya, 1995
	Increased by 5,5 times (to a greater degree - in the polluted areas) in comparison with the pre-Chernobyl level, Belarus	Manak et al., 1996
	Raised by 3.5 times in the Gomel area, 2.5 times - in the	Nesterenko, 1996

	Mogyliov area from 1986 level	
Early atherosclerosis and ischemic heart disease	Observed in evacuees and in heavy polluted districts in the Kiev area, Ukraine	Prokopenko, 2003
Leucopenia and anaemia	Increased by 7 times in comparison with 1985 levels in the Mogyliov area (Belarus) for the first three years after the catastrophe	Goffmann, 1994, p. 514
Infringement of the blood supply in legs	In girls who lived in the 137-Cs 1 - 5 Ci/km2 polluted areas of Belarus for 10 years	Khomitch, Lyusenko, 2002; Svanevsky, Gamshey, 2003
Changes of leukocytes number and activity	Lowered in pregnant women in the polluted territories of the Kursk area, Russia	Alymov et al., 2004
	Correlation between number of lymphocytes and basophilic cells with a level of 137-Cs pollution , Belarus	Miksha, Danilov, 1997
	Lowered in evacuees 7-8 years after the catastrophe, Ukraine	Baeva, Sokolenko, 1998

Table 9 Blood disease morbidity (per 100,000) in the adult population of Belarus, 1979 - 1997 (Gapanovich *et al.,* 2001)

	1979-1985	1986-1992	1993-1997
Acute leukemia	2,82±0,10	3,17±0,11*	2,92±0,10
Chronic leukemia	6,09±0,18	8,14±0,31*	8,11±0,26*
Eritremia	0,61±0,05	0,81±0,05*	0.98±0,05*
Multiple Mieloma	1,45±0,06	1,86±0,06*	2,19±0,14*
Hodgkins' disease	3,13±-0,10	3,48±0,12*	3,18±0,06
Non-Hodgkins' Lymphoma	2,85±0,08	4,09±0,16*	4.87±0,15*
Mielodisplastics syndrome	0,03±0,01	0,12±0,05*	0,82±0,16*

* P <0.05

7. Imbalance of hormone/endocrine statute

There is much evidence correlating endocrine/hormone imbalance with the radioactive Chernobyl fallout. Table 10 presents data concerning the incidence rate for Type 1 diabetes mellitus in Belarus.

Table 10 Incidence rate of Type 1 diabetes in children and adolescents in Belarus, 1980 – 2002 (Zalutskaya *et. al.,* 2004)

	1980-1986	1987 – 2002
Low polluted territories (Minsk area)	2,25 ±0,44	3,32 ± 0,49
Heavily polluted territories (Gomel area)	3,23 ± 0,33	7,86 ± 0,56*

* P <0.05

Table 11 presents some research showing correlations between the Chernobyl radioactive pollution with the development of hormone/endocrine diseases.

Table 11 Endocrine/hormonal diseases in some Chernobyl polluted territories

Disease	Area, feature	Author
Thyroid gland diseases (autoimmune thyroiditis, thyrotoxicosis, diabetes etc.)	Raised since 1992 in the polluted territories, Ukraine	Tron'ko et al.,, 1995
	Exceeds by 600 times the pre-Chernobyl level, Gomel area, Belarus	Byrjukova, Tulupova, 1994
	Increased among 1,026 046 recently confined women from territories with more than 1 Ci/km2, Belarus	Busuet et al., 2002
	Auto-immune thyroiditis in children increased by nearly 3 times during the 10 years after the catastrophe, Belarus	Leonova, Astakhova, 1998
	Congenital diabetes increased in Belarus (before the catastrophe there were no recorded cases)	Marples, 1996; Zalutskaya, *et al.,* 2004
	Children goiter increased by 7 times (since 1985) by 1993 in the Gomel area, Belarus,	Astachova, *et al.,* 1995
	In Ukraine, Belarus and Russia a survey of children showed diseases in 38,5 % (45,873 cases) in the 10 yrs following the accident	Yamashita, Shibata, 1997
	In 47 % (from 3437 surveyed children) in the Mozyrskiy district of Gomel area, Belarus	Vaskevich, Chernyshova, 1994
	In each second child in Bryansk area, Russia	Kashirina, 2005
	7,601 cases in children (among 284 cancer cases) in 1998 – 2004, Bryansk area, Russia	Karevskaya *et al.,* 2005

Endocrine system diseases	Increased 5 times by 2002 in children in the polluted districts of the Tula area, Russia	Sokolov, 2003
	Exceeded 2,6 times the average areas' level in the polluted southwest districts, adult population, Bryansk area, Russia	Sergeeva et al., 2005
Concentration of hormones	Many cases of hormonal imbalance in the first two years after the catastrophe in the heavily polluted territories, Ukraine	Stepanova, 1995
	Lowered testosterone level in the heavily polluted territories, Belarus	Lialykov et al., 1993
Concentration of hormones	Increase concentration of T4 and TCG and decreases - T3 hormones in polluted districts in Gomel and Vitebsk areas, Belarus	Dudynskaya, Suryna, 2001
	Increase of insulin level in boys and girls, and testosterone level - in girls in the polluted territories, Belarus	Antipkin, Arabskaya, 2001; Leonov, 2001
	Cortisol level increased in pregnant women parallel with the level of incorporated 137-Sr	Duda, Kharkevich, 1996
	Cortisol, thyroxin, progesterone level in children correlated with the level of radioactive pollution, Belarus	Sharapov, 2001
	Cortysol level raised in territories with 1 - 15 Ci/km2, and lowered in territories with a greater level of pollution in healthy newborns, Gomel and Mogyliov areas, Belarus	Danil'chik et al., 1996; Petrenko, et al., 1993
Prolactinemia	17,7 % of the surveyed women of reproductive age, in the radioactively polluted Russian territories	Strukov, 2003

In 1993 more than 40 % of the surveyed children in the Gomel area of Belarus had an enlarged thyroid gland. By expert estimations, in Belarus up to 1,5 million are at risk of a pathology of the thyroid gland (Goffmann, 1994a; Lipnick, 2004).

8. Immune disturbances

Additional irradiation, even at a small level, will attack the immune system. Results of further important research carried out over recent years in Ukraine, in Belarus and Russia is presented in Table 12

Table 12 Immunity disturbances in some of Chernobyls' polluted territories.

Disease	Feature, area	Author
Changes to cellular and humoral immunity	Decrease in immune response in healthy adults in the polluted territories of Belarus, and Russia	Soloshenko, 2002; Kyril'chik, 2000; Dubovaya, 2005
	5-fold increase of infringements of immunity and metabolism in children in polluted districts of the Tula area, Russia	Sokolov, 2003
	Children in the polluted territories, Russia	Terletzkaya, 2003
	In healthy children living in the polluted territories of Belarus	Soloshenko, 2002; Kyryl'chek, 2000;
Level of immunoglobulins	Changes in the level of immunoglobulins A, M and G at the onset of lactation in women from the territories polluted by Cs-137 5 Ci/km2 in the Gomel and Mogyliov areas, Belarus	Iskrytskiy, 1995; Zubovich et al., 1998
Maintenance T- and B- lymphocites	Decreased level in adults from the polluted areas of Belarus	Bandajevsky, 1999
Resistance to infections and other diseases	Lowered in the polluted areas of Belarus	Bortkevich et al., 1996
Lowered immune status	45 % of children in the polluted territories of Ukraine	Gurmanchuk et al., 1995
	Only 7 years after the catastrophe was there a normalization in children of a number of immune characteristics, Ukraine	Kharytonyk et al., 1996
	In babies in the territories with a level of pollution 5 and more Ci/km2 , Belarus	Petrova et al.,, 1993
Tonsillitis, lymphadenopaties	Raised frequency and expressiveness (45,4 % among the 468 children and teenagers	Bozhko, 2004

	surveyed), in the more polluted territories, Ukraine	
Anti-cancer immunity	Lowered in children in heavily polluted territories, and also among evacuees, Belarus	Nesterenko et al., 1993
Allergies Allergies (Cont'd)	More than twice the cows milk proteins among 1,313 Belarusian children in the territories 137Cs, 1 - 5 Ci/km2, than in the less polluted territories (36,8 % and 15,0 %).	Bandajevsky et al., 1995, 1995a; Bandajevsky, 1999

The typical consequence of infringement on the immune system in the radioactively polluted territories appears as an immuno-deficiency and, due to an increase in frequency and intensity of any acute and chronic diseases, is observed everywhere in the Chernobyl polluted territories. Sometimes the weakening of the immune system in these radioactively polluted territories is referred to as «Chernobyl AIDS».

9. Premature Ageing

There is accelerated ageing among the people in the Ukrainian radioactively polluted territories: their biological age exceeds their calendar one by 7 - 9 years (Mezhzherin, 1996). In territories polluted above 15 Ci/km2 on 137Cs in Belarus the mean age of men and women who died from heart attacks was 8 years younger than the average across Belarus (Antypova, Babichveskaya, 2001).

The array of diseases commonly considered exclusive to the elderly is typical for children in all of the heavily polluted territories. The immune system activity of these children is similar to the type of immune system activity experienced in old age. (Mezhzherin, 1996). The pathology of the digestive system epithelium in children from the polluted areas of Belarus also shows similarities with elderly people (Nesterenko, in litt.). Among 69 children and teenagers hospitalized in Belarus in 1991 - 1996 with alopecia, more than 70 % were from polluted territories (Morozevich et al., 1997)

10. Growing mutation rate

There are many studies showing a wide range of chromosomal aberrations in the Chernobyl radioactively polluted areas. A cytogenetic examination of inhabitants in the 30-km Zone shows that the frequency of aberrant cells and chromosomal aberrations for inhabitants in the Zone are significantly higher than those for the residents in the Kiev region, while the values of the latter group were found to be above the spontaneous and pre-Chernobyl levels (Table 13, Table 14).

Table 13 Comparative frequency (per 100 lymphocytes) of aberrant cells and chromosomal aberrations for Ukraine and the World

	aberrant cells	chromosomal aberrations	Authors
Ukraine, early 70-ies	Na	1.19±0.06	Bochkov et al., 1993
Ukraine, before 1986	1.43±0.16	1.47±0.19	Pilinska et al., 1999
The world generalized level, 2000	2.13±0.08	2.21±0.14	Bochkov et al., 2001
Ukraine, Kiev region, 1998-1999	3.20±0.84	3.51 ±0.97	Bezdrobna
30-km Zona, 1998-1999	5.02±1.95	5.32±2.10	et al., 2002

Table 14 Frequency (per 100 cells) of chromosomal aberrations in the Exclusion Zone inhabitants and residents in the Kiev area (Bezdrobna et al., 2002)

	30-km zone	Kiev area
Chromatid type	3.14 ± 0.24	2.33 ± 0.12
Fragments	1.59 ± 0.20	0.89 ± 0.12
Dicentrics + centric rings	0.33 ± 0.06	0.13 ± 0.03
Abnormal monocentrics	0.23 ± 0.05	0.12 ± 0.03
Total chromosome type	2.16 ± 0.24	1.18 ± 0.13

There is a growth in the level of some chromosomal aberrations in the 30-km Zone inhabitants over time (Table 15).

Table 15 Dynamics of frequency (per 100 cells) in different types chromosomal aberrations in the lymphocytes of the 30-km Zone residents (the same 20 persons) (Bezdobna et al., 2002)

	1998 - 1999	2001
Breaks	3.00 ± 0.33	3.53 ±0.40
Exchanges	0.16 ± 0.07	0.29 ± 0.07
Dicentric + centric rings	0.39 ± 0.09	0.45 ± 0.09
Total chromosome type	2.58 ± 0.35	1.63 ± 0.16 *

* $P < 0.05$

A comparison of the results of all the above mentioned cytogenetic examinations, together with many others presented in Table 16, allows us to assume the activation of mutagenesis in the regions with any level of the Chernobyl radioactive pollution.

19

Table 16 Examples of chromosomal aberrations and other genetic effects of the Chernobyl fallout

Character	Feature, area	Author
Frequency of chromosomal aberrations in somatic cells	Increase by 2-4 times in territories with pollution more than 3 Ci/km2, Ukraine, Russia	Bochkov, 1993
	Increase in the radioactively polluted territories, Belarus, Ukraine	Nesterenko, 1996, Goncharova, 2000; Maznik, Vinnikov, 2002; Maznik, 2003; Lyazuk *et al.*, 19944 Sevan'kaev *et al.*, 1995; Ivanenko *et al.*, 2004
	Correlates with the level of radioactive pollution in Ivano-Frankovsk area, Ukraine	Sluchick *et al.*, 2001
	More in thyroid cancer cells than in normal somatic cells in the polluted territories	Polonetskaya *et al.*, 2001
	More in children irradiated *in utero*, Ukraine	Stepanova, 1995; Stepanova *et al.*, 2002
	For children living in territories with 15 Ci/km2, unstable chromosome aberrations remained at a high level for 16 years; stable aberrations developed over the period	Pilinska *et al.*, 2003
Number of mitosis (mitotic index)	Lower in polluted districts of Bryansk area, Russia	Pelevina *et al.*, 1996
Mutations in satellite DNA	2-fold increase in children whose parents lived in the polluted territory of the Mogyliov region of Belarus ; correlated with the level of pollution, Belarus	Dubrova, Jeffreys, 1996; Dubrova, 2002; Dubrova *et al.*, 1996; 1997, 2002
Chromosomal aberrations and satellite DNA mutations	More in children with thyroid cancer	Melnov *et al.*, 1999
Chromosomal mutations de novo.	Increase in the polluted territories, Belarus	Lazjuk *et al.*, 2001

11. Infection and infestation

In the Chernobyl radioactively polluted territories (compared with clean territories) increasing morbidity by intestinal toxicosis, gastro-enteritis, dysbacteriosis, sepses, virus hepatitis and respiratory viruses was observed (Batyan, Kozharskaya, 1993; Kapytonova, Kryvitskaya, 1994; Nesterenko *et al.*, 1993; Kul'kova *et al.*, 1996; Busuet *et al.*, 2002; Antonov,Ostreiko, 1995; Goudkovsky *et al.*, 1995). Several examples of such studies are presented in Table 17.

Table 17 Examples of some infectious and parasitic diseases observed in the Chernobyl radioactively polluted territories

Disease	Feature, area	Author
Herpes virus infections	Active in the radioactively polluted territories, Belarus	Matveev *et al.*, 1993; Voropayev *et al.*, 1996
Trichocephalisis (Trichocephalis trichiurus)	Increased correlation with density of radioactive pollution in the Gomel and Mogyliov areas, Belarus	Stepanov, 1993
Pneumocistis (Pneumocystis carinii)	Increase in children in the polluted territories, Russia	Lavdovskaya *et al.*, 1996
Cryptosporidosis (Cryptosporidium parvum)	Raised in radioactively polluted territories in the Bryansk, Mogyliov and Gomel areas of Russia and Belarus	Lavdovskaya *et al.*, 1996
Tuberculosis (Mycobacterium tuberculosis)	Increased frequency and intensity in the polluted areas of Belarus.	Belookaya, 1993
	Occurrence of medicine-resistant forms and "rejuvenation" *{sic}* diseases in the polluted territories of Belarus	Borschevsky, 1996
Viral hepatits	2-fold increase (above Byelorussian mean level) in polluted territories of the Gomel and Mogyliov areas, during the 6 - 7 years after the catastrophe	Matveev, 1993
	An increase among adults and teenagers in the polluted territories of Vitebsk area, Belarus	Zhavoronok *et al.*, 1998
Cytomegalovirus (CMV) Infection (Cytomegalovirus hominis)	Active in pregnant women in the radioactively polluted territories, Belarus	Matveev, 1993

Microsporia (Microsporum sp.) occur in the radioactively polluted territories of the Bryansk areas (Russia) more frequently and in a more virulent form (Table 18).

Table 18 Microsporia cases (per 100 000) in three heavily radioactively polluted districts of the Bryansk area, Russia, 1998 - 2002 (Rudnytzkyi et al., 2003).

Years	Polluted districts	"Pure" districts
1998	56,3	32,8
1999	58,0	45,6
2000	68,2	52,9
2001	78,5	34,6
2002	64,8	23,7

12. Children's health

There are increases in children's general morbidity, and also increases in rare illnesses in the Chernobyl polluted territories of Ukraine, Belarus and Russia (Nesterenko et al., 1993).

In Belarus, unique inspections of the same children were conducted in 1995 - 1998 and in 1998 - 2001 (Arynchin *et al.,* 2002). One group (n - 133, 10,6 years) was from the heavily radio-contaminated territories, and the second (n - 186, 9,5 years - from low polluted territories. Results of the children's self-estimation of their health are presented in Table 19, and results of clinical inspection are presented in Table 20.

Table 19 Dynamics of children's complaints, from heavily and low irradiated territories of Belarus, (%) concerning their health (Arynchin *et al.,2002*)

	Heavy irradiated territories		Low irradiated territories	
	1995-1998	1998 - 2001	1995-1998	1998 - 2001
Total complains	72.2	78.9	45.7 **	66.1 *, ***
Weakness	31.6	28.6	11.9 **	24.7*
Dizziness	12.8	17.3	4.9 **	5.8 ***
Headache	37.6	45.1	20.7 **	25.9 ***
Syncope	0.8	2.3	0	0
Nasal bleeding	2.3	3.8	0.5	1.2
Fatigability	27.1	23.3	8.2 **	17.2*
Heart arrhythmia	1.5	18.8*	0 5	.8 *, ***
Stomachache	51.9	64.7*	21.2 **	44.3 *, ***
Belching	9.8	15.8	2.2 **	12.6*
Heartburn	1.5	7.5*	1.6	5.8*
Decreased appetite	9.0	14.3	1.1 **	10.3*
Allergic eruptions	1.5	3.0	0.5	5.8*

*P <0.05; ** P <0.05; *** P <0.05.

Table 20 Dynamics of clinical syndromes (%) and diagnoses for the same children (see Table 19) from the heavily and low irradiated territories of Belarus (Arynchin *et al.*, 2002) It is clear that children in heavily radio-polluted territories really do suffer, to a much greater degree, from a variety of diseases.

Clinical syndromes and diagnosis	Heavy irradiated territories		Low irradiated territories	
	1995 – 1998	1998 – 2001	1995 - 1998	1998 - 2001
Chronic gastritis	44.2	36.4	31.9	32.9
Chronic duodenitis	6.2	4.7	1.5	1.4
Chronic gastro-duodenitis	17.1	39.5*	11.6	28.7*
Bilious dyskinesia	43.4	34.1	17.4 **	12.6 ***
Vegeto-vascular and cardiac syndrome	67.9	73.7	40.3 **	52.2 ***
Astheno-neurotic syndrome	20.2	16.9	7.5 **	11.3
Chronic tonsillitis	11.1	9.2	13.6	17.2 ***
Caries	58.9	59.4	42.6 **	37.3 ***
Chronic periodontitis	6.8	2.4	0 **	0.6

*P <0.05; ** P <0.05; *** P <0.05.

Practically all forms of studied nosology are more prevalent in the main group compared to the control during both the first and the second examinations. Data of Table 19 and Table 20 give a convincing picture of sharply worsening health in children from the polluted territories.

Some examples of other investigations into children's health in the polluted territories are presented in the Table 21.

Table 21 Spectrum of children's non-cancer illnesses in the Chernobyl radioactively polluted territories

Disease	Feature, area	Author
Total child morbidity	Increase 2,9 times with 1986 on 2001, Ukraine	Moskalenko, 2003
	More intensive growth in the polluted territories of Kaluga area, Russia	Ignatov *et al.*, 2001
	2-fold increase above the mean in the Bryansk area in the polluted territories in 2004	Sergeeva *et al.*, 2005
Newborn morbidity	Annual growth in Belarus on 9,5 % ; the highest - in the most polluted Gomel area	Dzykovich *et al.*, 1996
The lowered weight of newborns	Irradiated *in utero*, Ukraine	Stepanova, 1995; Zakrevsky *et al.*, 1993
Circle of head	Less in newborns in the polluted territories of Ukraine and Belarus	Loganovsky, 2005; Akulich *et al.*, 1993
Infringements of rate of physical development	Irradiated *in utero*, Ukraine	Ushakov *et al.*, 1997
	In the polluted territories, Belarus	Sharapov, 2001
Premature birth	More often in the polluted territories, Belarus	Tsymlyakova, Lavrent'eva, 1996
invalided Children	351,9 (per 10 000), 1998 – 1999 in three polluted districts of the area (173,8 – in Bryansk area, 160, 7 - Russia)	Komogortseva, 2001
Rate of growth	Delay in the radioactively polluted territories, Belarus	Antipkin, Arabskaya, 2001
Cataract	Correlation between level of frequency and intensity within the radioactive pollution of territories, Belarus	Paramey *et al.*, 1993; Edwards, 1995; Goncharova, 2000;
	Correlation between frequency and level of incorporated 137Cs, Gomel area, Belarus	Bandajevsky, 1999
Retinal diseases	Occurs 6,2 % in 1985, 17,0 % in 1989 (4,797 children from two polluted districts of the Gomel area (137Cs up to 23 Ci/km2), Belarus.	Byrich et al., 1999

12.1. Children's respiratory system diseases

Respiratory system diseases occurred everywhere in the Chernobyl radioactively polluted territories (Table 22).

Table 22 Children's respiratory system diseases in the Chernobyl polluted territories

Respiratory syndrome	30 % of children in the polluted territories in the first months after the catastrophe, Ukraine	Stepanova, 1995
	Correlation frequencies with level of radioactive pollution in the Gomel area, Belarus.	Goudkovsky et al., 1995
	Children who at the moment of the catastrophe were 0 - 4 years, suffered more often in territories with 15 - 40 Ci/km2, than in territories with 5 - 15 Ci/km2.	Kul'kova et al., 1996
Suffocation (asphyxia)	In half of the 345 surveyed newborns irradiated *in utero*, Ukraine	Zakrevsky et al., 1993
Bronchial conductivity	In 53,6 % children in the polluted territories (surveyed more than 110 000) against 18,9 % on the low polluted territories, 1986 – 1987, Ukraine	Stepanov et al., 1995
Latent bronchospasm	In 69,1 % of children from the polluted territories (among more than 110, 000 surveyed) against 29,5 % in the lesser polluted territories, 1986 – 1987, Ukraine	
Bronchial asthma	Number of cases higher in the polluted territories, Belarus	Dzykovich et al., 1996; Sitnikov et al., 1993
	Heavier expression in the polluted territories, Russia	Terletskaya, 2003
Chronic bronchitis	Heavier expression in the polluted territories, Russia	
Nasopharyngeal chronic pathology	1.5 - 2 times more often in the polluted territories, Belarus	Dzykovich et al., 1996; Sitnikov et al.,, 1993
Acute respiratory diseases	2-fold in those irradiated *in utero*	Nesterenko, 1996

12.2. Children's Blood and lymphatic systems diseases

Cardiovascular system diseases in children occurred more frequently in the polluted territories. Table 23 presents data on the dynamics of reduction in the rate of blood formation in children after the catastrophe of Belarus.

Table 23 Reduction in the rate of blood formation in children from Belarus (Gapanovich et al., 2001)

	1979-1985	1986 – 1992	1993
The total cases of pre-leukemia conditions	65	98	78
Annual number of cases	9,3	14,0	15,6
The standardized parameter (per 10 000)	0,60±0,09	0,71±0,1*	
	1,00	1,46 *	1.73*

Data on children's diseases of the respiratory system in the radioactively polluted territories are presented in Table 24.

Table 24 Occurrence of children's blood and lymphatic system diseases in the Chernobyl polluted territories

Disease	Feature, area	Author
Infringements of cardiac rhythm	In more than 70 % of children with an average age of one year, in the territories 5 - 20 Ci/km2, Ukraine	Tsybul'skaya *et al.,* 1992
Infringements of vegetative regulation of cardiac activity	In children from the polluted territories of Belarus	Nedvetskaya, Lialykov, 1994; Sykorenskyi, Bagel', 1992; Goncharik, 1992
Arterial hypertension	Correlated with a level of the incorporated 137-Cs in Gomel area, Belarus	Kienya, Ermolitsky, 1997
Number of B- and T-- lymphocytes	Correlation with a level incorporated 137-Cs in the Mogyliov and Gomel areas, Belarus	Dzykovich *et al.,* 1996; Nesterenko, 1996;. Bandajevsky, 1999; Khmara *et al.,* 1993;
	In children 10 - 13 years old in heavily polluted territories of Kursk area, Russia	Alymov *et al.,* 2004
Lymphopenia	Increase of cases in the polluted territories, Belarus, Russia	Lukianova, Lenskaya, 1996; Sharapov,

		2001; Vasyna *et al.,* 2005
Bradycardia	More often in children and teenagers in the polluted territories, Ukraine	Kostenko, 2001
Lymphoid hyperplasia	Occurred in 30 % children in the heavily polluted territories in the first months after the catastrophe, Ukraine	Stepanova, 1995
Hematological disease	Occurred in 90 % of children in heavily polluted territories in the first months after the catastrophe, Ukraine	Stepanova, 1995
	Raised frequency among 1,220,424 newborns in the territories more than 1 Ci/km2, Belarus	Busuet *et al.,* 2002
	In 3,7 - 6,9 times above (compared to other areas) in the three most polluted districts of Bryansk area, Russia	Komogortseva, 2001
	Raised more than twice in 2002 in heavily polluted districts of the Tula area, Russia	Sokolov, 2000
Heart conductivity	The level of infringements correlated with radioactive pollution of territories, Belarus	Bandajevsky, 1997, 1999
Elasticity of arterial vessels	Declining in all children (including 'healthy' children) in the polluted territories of the Gomel and Brest areas, Belarus	Arynchin *et al.,* 1996

12.3. Dental system diseases

Dental diseases in children are more frequent in the Chernobyl radioactively polluted territories (Table 25).

Table 25 Occurrence of dental diseases in children in the Chernobyl polluted territories

Disease	Feature, area	Author
Caries	Raised prevalence and intensity in children and teenagers in more polluted territories, Belarus and Russia	Mel'nichenko, Cheshko, 1997; Sevbitov, 2005
Enamels acid resistance	Lowered in more polluted territories, Belarus	Mel'nichenko, Cheshko, 1997
Paradont pathologies	Greater intensity in those irradiated *in utero*, Russia	Sevbitov, 2005
Combinations of anomalies of dental systems	Greater intensity in those irradiated in utero; greater frequency in the polluted territories, Russia	Sevbitov, 2005

Frequencies of some dental diseases correlate with the level of radioactive pollution (Table 26).

Table 26 Frequency of dental anomalies (%) in children in territories with different levels of radioactive pollution (Tula and Bryansk areas, Russia), * (Sevbitov et al., 1999).

	Up to 5 Ci/km2	5-15 Ci/km2	15-45 Ci/km2	Children born
Teeth anomalies	3,7	2,4	2,8	Before 1986 (n=48)
	4,2	4,6	6,3	After 1986 (n=82)
Teeth row anomalies	0,6	0,4	0,6	Before 1986 (n=8)
	0,6	0,6	1,7	After 1986 (n=15)
Occlusion	2,6	2,4	2,2	Before 1986 (n=39)
	4,4	5,2	6,3	After 1986 (n=86)
Age norm	5.3	5,7	3,1	Before 1986 (n=77)
	2,6	2,0	0,6	After 1986 (n=28)

* 5 Ci/km2 - Don, the Tula area (183 persons); 5 - 15 Ci/km2 - Uzlovaya, Tula area (183 persons); 15 - 45 ¬ Ci/km 2 - Novozybkov, Bryansk area (178 persons).

12.4. Congenital malformations

Congenital malformations (CM), such as cleft lip and/or palate ("hare lip"), doubling of kidneys, polydactyly, anomalies in the development of nervous and blood systems, amelia (limb reduction defects), anencephaly, spina bifida, esophageal and anorectal atresia, multiple malformations etc., have increased in the Chernobyl radioactively polluted territories (Lazjuk *et al.,* 1997; 1999; Surykov, 1996; Goncharova, 2000, Ibragimova, 2003; Dubrova *et al.*, 1996, and many others).

Table 27 presents the CM frequency of children who were born 1 - 3 years after the catastrophe in 15 districts with more than 15 Ci/km2 by Cs-137 pollution, and in Table 27 - the generalized data across Belarus.

Table 27 Frequency of congenital malformations (per 1,000 born alive) in the 15 districts of Gomel and Mogyliov areas, Belarus, who were born before and after the Chernobyl catastrophe (Nesterenko, 1996)

	1982-1985	1987-1989
G o m e l a r e a		
Braginsky	4.09 ± 1.41	9.01±3.02
Buda-Koshelevsky	4.69 ± 1.21	9.33±2.03*
Vetkovsky	2.75 ± 1.04	9.86±2.72
Dobrushinsky	7.62 ± 1.96	12.58±2.55
El'sky	3.26 ± 1.35	6.41±2.42
Kormyansky	3.17 ± 1.20	5.90±2.08

Lel'tchitsky	3.28 ± 1.16	6.55±1.98
Loevsky	1.56 ± 1.10	3.71±2.14
Khoyniksky	4.37 ± 1.16	10.24±2.55*
Chechersky	0.97 ± 0.69	6.62±2.33*
M o g e l y o v a r e a		
Bykhovsky	4.00±1.07	6.45±1.61
Klymovichsky	4.77±1.44	3.20±1.43
Kostjukovichsky	3.00±1.22	11.95±2.88 **
Krasnopol'sky	3.33±1.49	7.58±2.85
Slavgorodsky	2.48±1.24	7.61±2.68
Cherykovsky	4.08±1.66	3.59±1.79
All territories	3.87±0.32	7.19±0.55 ***

* P>0.05; ** P>0.01; *** >P 0.001

Table 28 Occurrence of congenital malformations (per 1,000 born alive) in the territories with different levels of Chernobyl radioactive pollution, Belarus (Lazjuk *et al.,* 1999)

	≥15 Ci/km2, 17 districts		<1 Ci/km2, 30 districts	
	1982-1985	1987 - 1995	1982-1985	19987 – 1995
Anencephaly	0,28	0,44	0,35	0,49
Spina bifida	0,58	0,89	0,64	0,94*
Cleft lip and/or palate	0,63	0,94	0,50	0,95*
Polydactyly	0,10	1,02*	0,26	0,52*
Limb reduction defects	0,15	0,49*	0,20	0,20
Down's syndrome	0,91	0,84	0,63	0,92*
Multiple malformations	1,04	2,30*	1,18	1,61*
Total	3,87	7,07*	3,90	5,84*

*P>0.05

Table 29 presents data on the occurrence of genetic malformations in the polluted territories, such as Belarus, Russia and Ukraine (for other countries see paper by Inge Schmitz-Feuerhake in this book).

Table 29 Occurrence of congenital malformations (CM) in the Chernobyl polluted territories

Character	Feature, area	Author
Central Nervous system CM	Increase among newborns who have died between 1987 - 1995 in the polluted areas, Belarus	Dzykovich, 1996
	Increased frequency of dropsy of the brain -19,6 %, brain tumor (medulloblastoma) - 59,6 % from 1987 to 1992 , Ukraine	Orlov, 1993
All cases registered CM	Increase in legal abortions in the polluted areas, Belarus	Lazjuk et al., 1997
	Increase in the polluted territories of Bryansk and Tula areas, Russia	Ljaginskaja, Osypov, 1995; Ljaginskaja et al., 1996;
	2-fold increase among children irradiated between 4-6 months in utero, than irradiated earlier and later, Belarus	Lomat', 1996
	1,8 increase in the polluted territories between 1985-1999 in the Zhytomir area, Ukraine	Fedoryshin et al., 2001
	Increase annually up to 2,500 with a peak in 1990, Ukraine , Belarus	Golubchikov et al., 2002; Orlov, 1995; Goncharova, 2000; Lazjuk et al., 1997
Down's syndrome	Increase among those irradiated in utero, Ukraine	Stepanova, 1995
	Increase in the more polluted territories of Bryansk area, Russia	Kapustina, 2005
	Increase among abortions in the heavily polluted territories, Belarus	Ljaginskaja et al., 1997
	Rise of frequency in January 1987 in the Gomel and Minsk areas, Belarus	Lazjuk et al., 2002
Eye CM	4-fold increase in 1988 - 1989 compared to 1961 - 1972 in the Gomel area, Belarus	Byrich et al., 1999

12.5. Mental disorders

Disorders of intellectual development in children irradiated *in utero* and/or caused by irradiation in the polluted territories are the most tragic consequences of the Chernobyl catastrophe's impact on health (see also paper by Konstantin Loganovsky in this book). Irradiated children have not kept pace with other children (Table 30).

Table 30 Number of children (%) with revealed deviations of intellectual development in the radioactively polluted territories of Ukraine, Belarus and Russia
(Medical consequences..., 1995; Prilipko et al., 1995)

Group of children	Tests			Degree of nervous disorders	
	«Man Picture»	Raven color matrixes	«British dictionary»	Ratter Scale A(2)	Ratter scale B(2)
Belarus (n = 1861)					
Polluted	3.2	17.2	31.6	42.7	34.3
Clean	3.0	15.1	20.6	26.6	33.3
Russia (n=1025)					
Polluted	5.7	5.9	12.9	50.0	50.0
Clean	4.7	2.0	2.7	33.3	33.3
Ukraine (n=1347)					
Polluted	4.3	10.9	7.2	53.9	63.2
Clean	2.6	3.5	4.9	29.9	33.5

Children 6 - 7 years old born in May – February, 1987 whose mothers were evacuated from a zone more than 40 Ci/km2, or lived in a zone 5-40 Ci/km2   irradiated *in utero*, suffered a greater frequency of neurotic disorders, CNS pathology and delay of mental development than children in the not so polluted ("clean") areas of Belarus (Table 31).

Table 31 Influence of irradiation *in utero* on the children's intellectual development, Belarus
(Gaiduk *et al.*, 1994; Kolominsky, Igumnov, 1994)

Character	Irradiated (n - 154)	Control group (n - 90)
Neurotic disorders	36,40 ± 3,88	13,30 ± 3,58*
Asthenic syndrome	53.5	15.6**
Vegetative dystonia	76.6	33.3*
CNS organic pathology	20,80 ± 3,15	6,70 ± 2,63**
Delay of mental development	18,80 ± 3.15	7,80 ± 2,82*
EEG Pathology: slow type	31.2	2.2

*P> 0.01; ** P> 0.001.*

Some publications testifying to the Chernobyl irradiation impact on intellectual and mental development are briefly listed in Table 32.

Table 32 Summary data on the Chernobyl irradiation impact on psychological development of children

Character	Features, area	Author
Delay of speech ability, lowered psycho - emotional development, low IQ indices	In children irradiated *in utero* , Belarus, Ukraine	Belookaya, 1993; Bugaev *et al.*, 1993; 1995; Basyl'chik, Lobach, 1995; Nyagu, Loganovsky, 1998; Kolominsky *et al.*, 1999; Basyl'chik *et al.*, 2001; Nyagu *et al.*, 2002;
Deviations in mental development	In children 5-6 years old irradiated *in utero*, Ukraine	Pasechnik, Chuprykov, 1993;
	In almost 30 % irradiated *in utero,* Ukraine	Stepanova, 1995
	More frequent in those irradiated *in utero* (comparison of 895 children who were born 1984 – 1990) in Tula area, Russia	Ermolyna, Suchotina, 1995
delayed mental development, memory impairment, immaturity for school	irradiated *in utero*, Ukraine	Zapesochnyi *et al.*, 1995
Organic pathology of a brain	2 % of those irradiated *in utero* among evacuees, Ukraine	Nyagu, Loganovsky, 1998
Decreased psychomotor development	In those irradiated *in utero* at 8 - 25 weeks of pregnancy, Novozybkov city, Bryansk area, Russia	Ulyanova *et al.,* 1995
Character of neurological reactions	Distinctions between children 12-13 years old in the heavily and the lesser polluted territories of Bryansk area, Russia	Korsakov, 2005
Epilepsy and epilepsy-related condition	Increase in children evacuated from Pripyat city, Ukraine	Chuprykov *et al.*, 1996
Intellectual insufficiency	Children 7-9 years old, correlation with level of irradiation *in utero* , Belarus, Ukraine	Stepanov *et al.*, 1993; Igumnov *et al.*, 1993
Delay of 0,5 - 1,5 years in	Irradiated *in utero* (370 children, studied at the age	Tereschenko *et al.*, 1992

speech development; a delay of psychomotor development; decreased threshold of convulsive readiness, backlog of psychomotor development	of 3,5 - 5 years), Belarus	
Schizophrenia	Highest rate among children born in the polluted territories	Ermolyna., 1994
	Increase in all irradiated groups	Loganovsky, Loganovskaya, 2000

Official explanations of this deep infringement on the mental ability of irradiated children claim that it is the result of stress, not the impact of irradiation. However, stress does not lead to schizophrenia, epilepsy or organic damage of the brain.

13. Case study: Ukrainian district post-Chernobyl health

The above listed morbidity, by separate groups of illnesses, does not give us the full picture of the devastation in health occurring in the Chernobyl radioactively polluted territories. In an attempt to observe such a picture, Table 33 presents some health statistics for one remote Ukrainian administrative district from the Zhytomir area – Lugyny district. The Lugyny administrative district is situated 110 km South-West of Chernobyl NPP. The population in 1986 was 29,276, and in 1996 - 22,552, including 4,227 children. 22 villages are situated in this territory with 1-5 Ci/km2 and 26 villages situated in the territory with < 1 Ci/km2. All medical information for this study was collected by the same persons, with the same equipment and protocols before and after the catastrophe in Central District Hospital in Lugyny.

Table 33 Dynamics of Health Status of Residents of Lugyny District ten years after the Chernobyl catastrophe (Godlevsky, Nasvit, 1999).

	1984 - 1985	1995 - 1996
Remaining life after diagnosis of lung and stomach cancers (months)	38 – 62	2 – 7,2
Detection of the first phase of destructive tuberculoses (% to total tuberculosis first time detected, per 100 000)	17,2 – 28,7	41,7 – 50,0
Endocrine pathology (per 1000 children)	10,	90 – 97;
Goiters (per 1000 children)	not register	12 – 13
Neonatal morbidity (per 1 000):	25 – 75	330 – 340
Total mortality (per 1 000)	10.9	15,5
Life expectancy	75 years	65 years

Fig. 1 presents the annual rate of congenital malformation in the Lugyny district area.

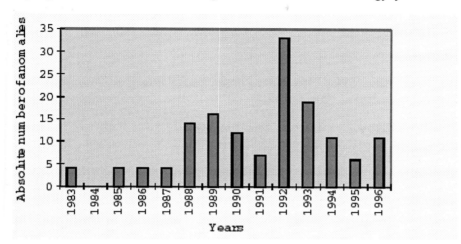

Fig. 1. Annual rate of congenital malformations in the Lugyny district area (Godlevsky, Nasvit, 1999).

I have reviewed above only a small part of the existing data on public health after the Chernobyl catastrophe. Examining data concerning the consequences for public health following the Chernobyl catastrophe does not give cause for optimism: mortality and morbidity in the polluted territories will continue to grow. This conclusion is based (among other arguments) on:

- A growing contribution to the collective dose in territories with a small density of radioactive pollution;
- Increasing (instead of decreasing) radioactive impact (dose and dose rate) due to increasing internal irradiation;
- The end of the latent period for several cancer diseases;
- Intensification of many non-cancer illnesses as a result of damage to the immune system and deepening of genetic instability.

Each year it has become clearer that the real consequences of this catastrophe are much more widespread and severe than has been predicted by the nuclear industry's adepts. In spite of this we, far from understanding the complete picture of the consequences of the Chernobyl catastrophe, are justified in saying that this catastrophe really is on a global scale and will continue to be so for many hundreds of years.

A better estimate is needed of the real health consequences of the catastrophe to mitigate these consequences as best we can. But, at minimum, the following is urgently required:

- An intensification and a broadening (instead of a reduction – as has happened over recent years in Russia, Ukraine and Belarus) of medical, ecological and radiological investigations;
- A strengthening and enrichment of the system of medical clinics and hospitals all over the polluted territories;

- An assessment for each irradiated person of his/her real individual dose and, from this base, create a lifespan plan of medical support for each person.
- Policy makers have to look the truth in eyes. Instead of the pro-Nuclear slogan "It's time to forget Chernobyl" another one is needed: "It's time to find ways to live with Chernobyl- forever".

References

Akulich N.S., Gerasymovich G.I. 1993. Indices of physical development of newborns after law dose irradiation. "Belarusian Children Health under modern ecological conditions (Consequences of the Chernobyl catastrophe)". VI Pediatric Congress of Belarus, Coll. Papers, Minsk, p.9 (in Russian).

Alexievich S. 1997. Chernobyl prayer (chronic for future). Moscow, Publ. "Ostozh'e", 223 p. (in Russian).

Alymov N.I., Pavlov A.Yu., Sedunov S.G., Gorshenin A.V., Popovich V.I., Loskutova N.D., Belobrovkin E.A. 2004. Immune system at person living on the Chernobyl irradiated territories. Russ. Sci.- Pract. Conf.: «Med.- Biol. Problems of Radio- and Chemical Protection., 20 - 21 May, 2004, Sant-Peterburg, Russia", , Coll. Papers, Sank-Peterburg, pp. 45 – 46 (in Russian).

Antipkin Yu.G., Arabskaya L.P. 2001. 3rd Int. Conf. "Medical consequences of the Chernobyl catastrophe: the results of 15 years investigations", 4 - 8 June, Kiev, Ukraine", Abstracts, Kiev, pp. 151 – 152 (in Russian).

Antonov E.Z., Ostreiko N.N. 1995. Lymph-adenoid ring' condition and children' enterobiosis in Gomel. Int. Sci. Conf. devoted by 5-th years establishing of the State Gomel Med. Ins., 9 – 10 November, 1995, Gomel, Belarus", Materials, Gomel, pp. 86 – 87 (in Russian).

Antypova S.I. Babichevskaya A.I. 2001. Belarusian adult mortality of the evacuees. 3rd Int. Conf. "Medical consequences of the Chernobyl catastrophe: the results of 15-years Researches, 4 - 8 June, 2001, Kiev, Ukraine", Abstracts, Kiev, pp. 152 – 153 (in Russian).

Arynchin A.N., Krotkaya N.A., Bortnik O.M. 1996. Brain circulation at children living on the radioactive contaminated territories of Belarus. Inter. Sci. Conf.: "10 years after the Chernobyl catastrophe (scientific aspects of problem). 28 - 29 February 1996, Minsk, Belarus", Abstracts, Minsk, p. 13 (in Russian).

Arynchin A.N., Avhacheva T.V., Gres N.A., Slobozhanina E.I. 2002. Health State of Belarusian Children suffering from the Chernobyl Accident: Sixteen Years after the Catastrophe. In: Imanaka T. (Ed.). Recent Research Activities about the Chernobyl Accident in Belarus, Ukraine and Russia. Res. Reactor Institute, Kyoto University (KURRI-KR-79), Kyoto, pp. 231- 240.

Astachova L.N., Demidchik E.P., Polyanskaya O.N. 1995. The main radiation risk' system: Thyroid carcinoma at Belarusian children after the Chernobyl accident. IV Int. Conf. : "Chernobyl catastrophe: Prognosis, prophylactics, treatment and medical-psychological rehabilitation of suffers." Collect. Materials., Minsk, pp. 119 — 127 (in Russian).

Babych T., Lypchanska L.F. 1994. State of hyphophisis – thyroid system women under low radiation impact. "Functional methods in obstetrician and gynecology", Sci.-Pract. Conf. Ukrainian obstetr.-gynecologists, 19—20 May, 1994, Donetsk, Ukraine", Abstracts, Donetsk, p. 9 (in Ukrainian).

Baeva E.V., Sokolenko V.L. 1998. T-lymphocyte's surface marker's expression after low dose irradiation. Immunol., № 3, pp. 56 - 59 (in Russian).

Basyl'chik S.V., Lobach I.V. 1995. Children's intellectual development after ionizing radiation of radio-Iodine nuclides in utero and in the first year of life with connection of the Chernobyl accident. Sci. Conf.: "Actual and prognosis disturbances of the psychic health after nuclear catastrophe in Chernobyl, 24 - 28 May, 1995, Kiev, Ukraine". Materials, Kiev, Assoc. "Chernobyl Doctors", p. 306 (in Russian).

Bandajevsky Yu.I., Kapytonova E.K., Troyn E.I. 1995. Expression of allergy to cow milk proteins and cortisol level in blood at children from the radionuclides polluted territories. In: "Actual problems of Immunology and Allergology.", III Congress Belarus. Sci. Soc. Immun. Allerg., Abstracts, Grodno, p. 111 (in Russian).

Bandajevsky Yu.I. 1999. Pathology of incorporated ionizing radiation. Gomel State Med. Inst. Minsk, 136 p. (in Russian).

Batyan G.M., Kozharskaya L.G. 1993. Juvenile rheumatoid arthritis at children from the radioactively polluted areas. VI Byelorussian Pediatric Congr.: "Byelorussian children' health under modern ecological condition (Consequences of the Chernobyl catastrophe)." Materials, Minsk, pp. 18—19 (in Russian).

Belookaya T.V., Koryt'ko S.S., Mel'nov S.B. 2002. Medical effects of the low doses of ionizing radiation. 4th Int. Congress on Integrative Anthropology, Materials, Sant-Peterburg, pp.24 - 25 (in Russian).

Belookaya T. V. 1993. Dynamics of the children health status of Belarus under modern ecological conditions. "The Chernobyl catastrophe: diagnostics and medical-psychological rehabilitation of the suffers. Collection of Conf. Materials , Minsk, pp. 3-10 (in Russian).

Bezdrobna L., Tsyaganok T., Romanova O., Tarasenko L., Tryshyn V., Klimkina L. 2002. Chromosomal Aberrations in Blood Lymphocytes of the Residents of 30-km Chernobyl NPP Exclusion Zone. In: Imanaka T. (Ed.). Recent Research Activities about the Chernobyl NPP Accadent in Belarus, Ukraine and Russia. Kyoto Univ. Res. Reactor Inst. (KURRI-KR-79), Kyoto, pp. 277 – 287.

Byrich T.V., Birich T.A., Pesaerenko D.K. 1994. Diagnostics, clinical characters and prophylactics of the cancer' setback at adults and children. Conf.: "The Chernobyl catastrophe: prognosis, prophylactics, treatment and medical-psychological rehabilitation of the suffers. Collect. of Materials, Minsk, pp. 32 – 34 (in Russian).

Byrich T.A., Chekina A.Yu., Marchenko L.N., Ivanova V.F., Dulub L.V. 1999. Ophtalmo-pathology at children are inhabit of the radioactively polluted territories of Belarus, and liquidators. Ecological Anthropology. Almanac. Minsk, Byelorussian Committee "Chernobyl Children", pp. 183 – 184 (in Russian).

Byrjukova L.V., Tulupova M.I. 1994. Dynamics of the endocrine pathologies in Gomel area, 1995—1993. Materials Int. Sci. Symp. «Medical aspects of the radioactive impact on population in the Chernobyl polluted territories», Gomel, p. 29 (in Russian).

Bochkov N.P. 1993. Analytical review of cytogenetic studies after Chernobyl accident. Russ. Med. Acad. Herald, # 6, cc. 51-56 (in Russian).

Bochkov N.P, Chebotarev A.N.,.Katosova L.D, Platonova V.I.. 2001. Data base for quantitative characteristics of chromosomal aberrations frequency in human peripheral blood lymphocytes assay. Genetic, vol. 37, N 4, pp. 549 - 557 (in Russian).

Bogdanovich I.P. 1997. Comparative analysis of children (0-5 years) mortality in 1994 in the radioactively polluted and clean areas of Belarus. Medical;-biolog. Effects and ways to overcomes the consequences of the Chernobyl accident. Collection of sci.

papers devoted by 10th anniversary of Chernobyl accident, Minsk –Vitebsk, p. 4 (in Russian).

Bozhko A.V. 2004. Remote results of long-time impact of low ionizing irradiation on lymphoid structures of children pharynx. Otolaryngology Herald , № 4, pp. 9-10 (in Russian).

Bortkevich L.G., Konoplya E.F., Rozhkova Z.A. 1996. Immunotropic effects of the Chernobyl catastrophe. Conf. "10 years after the Chernobyl catastrophe: Scientific problems", Abstracts, Minsk p. 40 (in Russian).

Borschevsky V.V., Kalechitc O.M., Bogomazova A.V. 1996. Tendencies in the tuberculosis morbidity after the Chernobyl catastrophe in Belarus. Med.-biol. aspects of the Chernobyl accident, № 1, pp. 33-37 (in Russian).

Bugaev V.N., Lagutin A.Yu., Druzhynyna E.S. 1993. Sensor's speech infringements and vegeto-distonia at 4-5 years old children living in Kiev. Social-psychological and psycho-neurological aspects of the Chernobyl accidents consequences. Materials of the scientific Conf. of the CIS states with international participation. Kiev, p. 229 (in Russian).

Bugaev V.N., Pyatak O.A., Lagutin A.G., Shul'zhenko V.B. 1995. Psychical health of the irradiated Ukrainian children. Materials Inter. Conf. "Actual and prognostic infringements of psychic health after nuclear catastrophe in Chernobyl", 24 – 28 May, 1995, Ukraine, Kiev., p.18 (in Russian).

Chuprykov A.P., Chupykova E.G., Danylov V.M., Myljuta E.L. 1996. Phenomenon of paroxysmal braking at children after the Chernobyl accident. Arch. Psych., № 10 – 11, pp. 53 - 54 (in Ukraine).

Danil'chik V.S., Ustynovich A.K., Vasylevsky I.V. 1996. Hormonal- biochemical homeostasis at the newborns in the radioactively polluted areas. Public Health, № 5, pp. 17-19 (in Russian).

Demedchik E.P., Demedchik Yu.E., Rebeko V. Ya. 1994. Thyroid' cancer in children. Materials Inter. Sci. Sympos. «Medical aspects radioactive impact on population in the Chernobyl polluted areas». Gomel, pp. 43—44 (in Russian).

Dubovaya N.I. 2005. Comparative analysis of allergosis' distribution among green-houses workers of Bryansk area. MaterilasInter. Sci.-pract. Conf. «Chernobyl – 20 years after. Social-Economical Problems and persective for development of suffering territories», Braynsk, p. 156 (in Russian).

Dubrova Y.E., Nesterov V.N., Krouchinsky N.G., Ostapenko V.A., Neumann R., Neil D.L., Jeffreys A.J. 1996. Human minisatellite mutation rate after the Chernobyl accident. Nature, vol. 380, April 15, pp. 683-686.

Dubrova Y.E., Nesterov V.N., Krouchinsky N.G., Ostapenko V.A., Vergnaud G., Giraudeau F., Buard J., Jeffreys A.J. 1997. Further evidence for elevated human minisatellite mutation rate in Belarus eight years after the Chernobyl accident. Mutat. Res., Vol. 381, pp. 267-278.

Dubrova Y.E., Grant G., Chumak A.A., Stezhka V.A., Karakasian A.N. 2002. Elevated Minisatellite Mutation Rate in the Post-Chernobyl Families from Ukraine. Am. J. Hum. Genet., vol. 71, pp. 800- 809.

Dubrova, Y.E. and Plumb, M.A. 2002. Ionising radiation and mutation induction at mouse minisatellite loci: The story of the two generations. Mutat. Res.,# 499, pp.143–150.

Duda V.I., Kharkevich O.N. 1996. Endocrine mechanisms of adaptations in the dynamics of gestation process at women under chronic radiation stress. «Motherhood and childhoods protection under impact of consequences of the Chernobyl catastrophe». Materials of scientific investigations, 1991-1995. Minsk, part. 1, pp. 96 – 99 (in Russian).

Dudynskaya R.A., Surina N.V. 2001. Condition of the thyroid system of родильниц, from the radionuclides' pollution in Gomel area. 3rd Iner. Conf. " Medical consequencies of the Chernobyl catastrophe: Results of 15-years investigations. 4-8 June 2001 , Kiev, Ukraine". Abstarcts, pp. 192 -193 (in Russian).

Dzykovich I.B., Kornilova T.I., Кот T.I., Vanilovich I.A. 1996. Health condition of the pregnant and newborns from different areas of Belarus. Medical-Biological aspects the Chernobyl accident. № 1, pp. 16 – 23 (in Russian).

Dzikovich I.V. 1996. Epidemiological analysis of ecologically dependent perinatal illnesses in Belarus. Sci. Congr.: "10 years after the Chernobyl catastrophe (scientific aspects of problem)" , Abstracts, Minsk, p. 87 (in Russian).

Edwards R. 1995. Will it get any worse? New Sci. Dec. 9. pp. 14 — 15.

Ermolyna L.A., Sosyukalo O.D., Suchotina N.K. 1994. Low doses impact on neuro-psychical health of children (methods and preliminary results). Part 1. Social and Clinical Psychiatry, vol 4, # 1, pp. 37 - 43 (in Russian).

Ermolyna L.A., Suchotina N.K. 1995. Comparative analysis of the neuro-psychical pathologies of group of children, irradiated in utero and in postnatal period. Materials Inter. Conf. "Actual and prognostic infringements of psychic health after nuclear catastrophe in Chernobyl. 24 – 28 May 1995. Kiev, Ukraine", Kiev, p. 310 (in Russian).

Evetz L.V., Lyalykov S.A., Ruksha T.V. 1993. Characters of immune infringements at children in connection of spectrum of the isotopic pollution of territory. Conf. "Chernobyl catastrophe: diagnostics and medical-psychological rehabilitation of suffers". Collection of Materials. Minsk, pp. 83 – 85 (in Russian).

Gaiduk F.M., Igumnov S.A., Shal'kevich V.B. 1994. Complex estimation of the psychical development of children suffering from radionuclides after the Chernobyl catastrophe. Social and clinical psychiatry, vol. 4, № 1, pp.. 44 – 99 (in Russian).

Gapanovich V.M., Shuvaeva L.P., Vinokurova G.G., Shapovalyuk N.K., Yaroshevich R.F., Mel'chakova N.M. 2001. Impact of the Chernobyl catastrophe to blood' depressions at Belarusian children. 3rd Iner. Conf. "Medical consequencies of the Chernobyl catastrophe:results of 15-years investigations. 4-8 June, 2001, Kiev, Ukraine", Abstracts, pp. 175 -176 (in Russian).

Godlevsky I., Nasvit O. 1999. Dynamics of Health Status of Residents in the Lugyny District after the Accident at the ChNPS. In: Imanaka T. (Ed.) Recent Research Activities about the Chernobyl NPP Accident in Belarus, Ukraine and Russia. Kioto Univ. Res. Reactor Inst. (KURRI-KR-79, pp. 149 – 157.

Golovko O.V., Izhevsky P.V. 1996. Studies of the reproductive behaviour in Russian and Belorussian populations, under impact of the Chernobyl ionizing irradiation. Rad. Biol., Radioecol., vol. 36, # 1, pp. 3 – 8 (in Russian).

Golubchikov M.V., Michnenko Yu.A., Babynets A.T. 2002. Changes in the Ukrainian public health at post-Chernobyl period. 5 Annual Sci.-Pract. Conf.: "To XXI Century - with safety nuclear technologies", Slavutich, 12-14 January, 2001". Scientific and Technological Aspects of Chernobyl, № 4, pp. 579 – 581 (in Ukrainian).

Goncharik I.I. 1992. Arterial hypertension at inhabitant's Near-to-Chernobyl zone. Belarusian Public Health, № 6, pp. 10 – 12 (in Russian).

Goncharova R.I. 2000. Remote consequences of the Chernobyl Disaster: Assesnment After 13 Years. In: E.B. Burlakova (Ed.). Low Doses Radiation: Are they Dangerous? NOVA Sci. Publ., pp. 289 – 314.

GorptchenkoI.I., Ivanyuta L.I., Sol'sky Ya.P. 1995. Genital System. In: Baryachtar V. G. В.Г.(Ed.). Chernobyl Catastrophe. Kiev, «Naukova Dumka» Publ., pp. 471 – 473 (in Russian).

Goffman J. 1994a. Chernobyl accident: radioactive consequences for the existing and the future generations. Minsk, «Vysheihsaya Shkola» Publ., 576 p. (in Russian).

Goffman J. 1994b. Radiation-Induced Cancer from Low-Dose Exposure: an Independent Analysis. Translation from English Edition (1990). Social Ecological Union Publ., Moscow, vol. 1,2, 469 p. (in Russian).

Grodzinsky D.M. 1999. General situation of the Radiological Consequences of the Chernobyl Accident in Ukraine. In: Imanaka T. (Ed.) Recent Research Activities about the Chernobyl NPP Accadent in Belarus, Ukraine and Russia. Kyoto Univ. Res. Reactor Inst. (KURRI-KR-7), pp. 18 – 28.

Gudkovskyi I.A., Kul'kova L.V., Blet'ko T.V., Nechai E.V. 1995. Children health and level of Cs-137 pollution in the invalitated territories. Intern, Sci. Conf. Devoted 5th anniversary of Gomel State Med. Inst., 9 - 10 November 1995, Гомель, Belarus", Materials, Gomel, pp. 12 – 13 (in Russian).

Gurmanchuk I.E., Tythov L.P., Kharytonik G.D., Kozlova N.A. 1995. Comparative characteristics of immune status of the sick children in Gomel', Mogilyov' and Brest' areas. Actual problems of immunology and allergology. 3rd Congress Beloru. Sci. Society of Immunologists and Allergologists. Abstracts. Grodno, pp. 79 – 80 (in Russian).

Health Consequences of the Chernobyl Accident. 1995. Results of the IPHECA Pilot Projects and Related National Programmes. Scientifical Report of the International Programme on the Health Effects of the Chernobyl Accident. WHO. Geneva, 560

Ibragimova A.I. 2003. Clinical data on genotoxic action of ionizing radiation. Russ. Herald Perinatol., Pediatr., vol. 48, № 6, pp. 51 – 55 (in Russian).

Ivanenko G.F., Suskov I.I., Burlakova E.B. 2004. Glutathione level and cytogenetic characters in peripheral lymphocytes at children under low dose impact. Herald of Russ. Acad. Sci., Seria Biologia, № 4, pp. 410 – 415 (in Russian).

Ignatov V.A., Selyvestrova O.Yu., Tsurykov I.F. 2001. Eho 15th post-Chernobyl years in Kaluga land. "Legacy of Chernobyl". Materials Sci.-Pract. Conf. "Medical-psychological , radio ecological and social-economical consequences of the Chernobyl accident in Kaluga area". Kaluga, Obninsk, #   3, pp. 6 – 15 (in Russian).

Igumnov S.A., Sekatch N.S., Chuiko Z.A. 1993. Complex diagnostics of psychic development of children after of radioactive impact during critical period of cerebrogenesis. Conf.: "Chernobyl catastrophe: diagnostics and medical-psychological rehabilitation of the suffers". Collection of materials, Minsk, pp. 14 — 15 (in Russian).

Imanaka T. (Ed.) 2002. Recent Research Activities about the Chernobyl NPP Accident in Belarus, Ukraine and Russia, Kyoto Univ. Res. Reactor Inst. (KURRI-KR-79), Kyoto, 297 p.

Iskrytskyi A.M. 1995. Humoral immunity and immunological characters of human milk in the radioactively polluted areas of Belarus. «Actual problems of immunology and allergology», III Congress Belorus. Sci. Soc. Immunologists and Allergologists. Abstracts, Grodno, pp. 85 – 86 (in Russian).

Ivanov V.K., Matveenko E.G., Byryukov A.P. 1996. Analysis of new registered illnesses at Kaluga area's liquidators. «Legacy of Chernobyl», Materials Sci.-Pract. Conf. "Medical-psychological , radio ecological and social-economical consequences of the Chernobyl accident in Kaluga area". Kaluga, Obninsk, # 2, pp. 233-234 (in Russian).

Ivanov V.K., Tsyb A.F., Nilova E.V. 1997. Cancer risks in the Kalyga oblast of the Russian Federation 10 years after the Chernobyl accident. Radiat. Environ. Biophys. Vol. 36, pp. 161-167.

Kapustina N.K. 2005. Dinamics of Congenital malformations in Bryansk area, 1999 - 2004. Materials Inter. Sci.-Pract. Conf.: «Chernobyl - 20 years after. Social-economical problems and perspective for development of the impacted territories», Bryansk, pp. 163 – 164 (in Russian).

Kapytonova E.K., Kryvitskaya L.V. 1994. Babies morbidity structure on the radioactively polluted territories in 6 years after Chernobyl accident. Materials Inter. Sci. Symp. «Medical Aspects of radioactive impact on populations after the Chernobyl accident», Gomel, p .52 (in Russian).

Karevskaya I.V., Kurbatskaya G.Ya., Vasil'tsova O.A., Stepunin L.A., Zubareva I.A. 2005. Dispanserization' role in the diagnostics of thyroids deceases at population of South-Western districts of Bryansk area. Inter. Sci.-Pract. Conf.: «Chernobyl - 20 years after. Social-economical problems and perspective for development of the impacted territories», Bryansk, Materials, pp. 164 – 165 (in Russian).

Kashirina M.A. 2005. Social-ecological factors of public health in the radioactively polluted territories of Bryansk area. Inter. Sci.-Pract. Conf.: «Chernobyl - 20 years after. Social-economical problems and perspective for development of the impacted territories», Bryansk, Materials, pp. 166 – 167 (in Russian).

Kharytonik G.D., Tytov L.P., Gurmanchuk I.E., Ignatenko S.I. 1996. Character and dynamics immunological characters at children living during seven years in the condition "clean" territories of Bragin district. Sci. – Pract. Conf. "Remote consequences of irradiation for immune and haemopoetic systems. 7 - 10 October 1996, Kiev, Ukraine", Abstracts, Kiev, pp. 59 – 60 (In Ukranian).

Khomich G.E., Lysenko Yu.V. 2002. Blood vessels' reographic characters of legs after change of spase position at girls with increasing vessel tonus and living in the radioactively polluted zone. Brest State Univ., Brest, 6 p. (in Russian).

Khmara I.M., Astakhova L.N., Leonova L.L. et al. 1993. Indices of immunity in children suffering with autoimmune thyroiditis. J. Immunol. № 2, pp. 56-58.

Kienya A.I., Ermolitskyi N.M. 1997. Vegetative component of child organism' activity with different level of incorporated Cs-137. In: Bandajevsky Yu.I. (Ed.) Structural and functional effects of incorporated radionuclides, Gomel, pp. 61 – 82 (in Russian).

Kolominsky Ya.L., Igumnov S.A. 1994. Social-psychological factors impact on psychical development of children 6-7 years old, from the Chernobyl territories. Inter. Conf. "Social-psychological rehabilitation of people suffering from ecological and thechnogenic catastrophes" Abstractes, Gomel, p. 33 (in Russian).

Kolominsky Y., Igumonov S., Drozdovitch V. 1999. The psychological development of children from Belarus exposed in the prenatal period to radiation from the Chernobyl atomic power plant. J. Child Psychol., Psychiat. Vol. 40, № 2, pp. 299-305.

Комogortseva L.K. 2001. Report for Bryansk Duma . prepared by representative L. K. Komogortseva on base on data from Bryansk Committee of Public Health and Bryansk Statistical Bureau. 31 января 2001 г., MS, 4 p. (in Russian).

Korsakov A.V. 2005. Comparisons of circulatory and neurological reactions of children from the ecologicvally, radioactively and toxically unsafe territories. Inter. Sci.-Pract. Conf.: «Chernobyl - 20 years after. Social-economical problems and perspective for development of the impacted territories», Bryansk, Materials, pp. 171- 173 (in Russian).

Krapyvin N.N. 1997. Chernobyl in Lypetsk: yesterday, today and tomorrow ... Lypetsk, 36 p. (in Russian).

Kul'kova L.V., Ispenkov E.A., Gutkovsky I.A., Voinov I.N., Ulanovcskaya E.V., Skidanenko G.I., Maevsky G.A., Yunhel' V.V. 1996. Epidemiological monitoring children health on the radionuclides' polluted territories of Gomel area. Med. Radiol. And Radioact. Safety, № 2, pp. 12 – 15 (in Russian).

Kulakov V.I., Sokur A.L., Volobuev A.L. et al. 1993. Female reproductive functions in areas affected by radiation after the Chernobyl power station accident. Environ. Health Perspect. Suppl., Vol. 101.

Kyra E.F., Tsvelev Yu.V., Greben'kov S.V., Gubin V.A., Chernichenko I.I. 2003. Woman reproductive health in the radioactively polluted territories. Military-Medical J., vol. 324, № 4, pp. 13 – 16 (in Russian).

Kyril'chek E.Yu. 2000. Characters of immune status and immune-rehabilitation of children on the radionuclides' polluted territories (clinic-laboratory investigation, 1996-1999). Thesis, Dr. Med., Minsk State Med. Inst., Minsk, 21 p. (in Russian).

Lavdovskaya M.V., Lysenko A.Ya., Basova E.N., Lozovaya G.A., Baleva L.S., Rybalkyna T.N. 1996. Ionizing radiation' impact on the distribution of cryptosporidiosis and pneumocystosis. Phraseology, № 2, pp. 153 – 157 (in Russian).

Lazjuk G.I., Kirillova I.A., Nikolaev D.L. 1994. Dynamics of congenital pathologies in Belarus and the Chernobyl catastrophe. Conf.: Chernobyl catastrophe: diagnostics and medical-psychological rehabilitation of the suffers". Collection of materials, Minsk, pp. 167 - 183 (in Russian).

Lazjuk G.I., Nikolaev D.L., Novikova I.V. 1997. Changes in registred congenital anomalies in the Republic of Belarus after the Cghernobyl accident. Stem Cells, 15, Suppl. 2, pp. 255 – 260.

Lazjuk G.I., Nikolaev D.L., Novikova I.V., Polityco A.D., Khmel R.D. 1999. Belarusian population radiation exposure after Chernobyl accident and congenital malformations dynamics. Int. J. Rad. Med. # 1, pp. 63-70.

Lazjuk G., Satow Y., Nikolaev D., Novikova I. 1999. Genetic Consequences of the Chernobyl Accident for Belarus Republic. In: Imanaka T. (Ed.) Recent Research Activities about the Chernobyl NPP Accident in Belarus, Ukraine and Russia. Kyoto Univ. Res. Reactor Inst. (KURRI-KR-7), pp. 174 - 177.

Lazjuk G.I., Naumchik I.V., Rumyantseva N.V., Polytyko A.D., Khmel' R.D., Egorova T.M., Kravchuk Zh.P., Verje P., Robert V., Satov Yu. 2001. Main results of studied of genetical consequences of the Chernobyl catastrophe at Belarus population. 3rd Inern. Conf. " Medical consequences the Chernobyl catastrophe: results of 15-years investigations. 4-8 June 2001, Kiev, Ukraine". Abstarcts, pp. 221 – 222 (in Russian).

Lazjuk G.I., Zatsepin I.O., Verje P., Ganier B., Robert E., Kravchuk Zh.P., Khmel' R.D. 2002. Down' Syndrom and ionizing radiation: reason-effect or bychance connetions. Rad. Biol., Radioecol., vol. 42, № 6, pp. 678 – 683 (in Russian).

Leonova T.A., Astakhova L.N. 1998. Autoimmune thyroiditis at pubertal girls. Public Health. № 5, pp. 30 – 33 (in Russian).

Ljaginskaja A.M., Osipov V.A. 1995. Comparison of estimation of reproductive health of population from contaminated territories of Bryansk and Ryazan areas of the Russian Federation. In: "Radioecological, Medical and Socio-economical Consequences of the Chernobyl Accident. Rehabilitation of Territories and Populations", Moscow. Abstracts, p. 91 (in Russian)..

Ljaginskaja A.M., Izhewskij P.V., Golovko O.V., 1996. The estimate reproductive health status of population exposed in low doses in result of Chernobyl disaster,. In: IRPA 9. Proceedings of the International Congress on Radiation Protection, Volume 2. pp. 62 – 67.

Ljaginskaja A.M., Golovko O.V, Osipov V.A. et al. 1997. Criteria for estimation of early deterministic effects in exposed populations, In: Third Congress of Radiation Research, Moscow. pp. 73-74 (in Russian).

Lialykov S.A., Evets E.V., Makarchik A.V. 1993. Peculiarity of the endocrine status of the children after long-time low doses irradiation. Conf.: "Chernobyl catastrophe: diagnostics and medical-psychological rehabilitation of suffers". Collection of Materials, Minsk, pp.. 68 — 70 (in Russian).

Lipnik B. 2004. The Earth and Radiation: realyty terrifyer than numbers.. (http://www.pravda.ru/science/planet/environment/47214-0)

Lisianyi N.I., Ljubich L.D. 2001. Role of the neuro-immune reactions for development of the post-radiation encephalopathy after low dose impact. 3rd Inter. Conf. "Medical consequences of the Chernobyl catastrophe: results of 15-years investigations, 4-8 июня 2001, Kiev, Ukraine." Abstracts, Kiev, p. 225 (in Russian).

Loganovsky K.N. 1998. Schizophrenic characters at persons after ionizing irradiation as result of the Chernobyl catastrophe. 2nd Inter. Conf. " Remote medical consequences of the Chernobyl catastrophe, 1– 6 June 1998, Kiev, Ukraine", Kiev, Abstracts, p. 78 (in Russian).

Loganovsky K.N. 1999. Clinic- epidemiological aspects psychiatric consequences of the Chernobyl catastrophe. Social and clinical psychiatry, vol. 9, # 1, pp. 5 – 17 (in Russian).

Loganovsky K. 2005. Health of children irradiated in utero.. (http://stopatom.slavutich.kiev.ua/2-2-7.htm).

Loganovsky K.N., Loganovskaya T.K. 2000. Schizophrenia spectrum disorders in persons exposed to ionizing radiation as a result of the Chernobyl accident. Schizophr. Bull., Vol. 26, pp. 751-773.

Lomat' L.N. Antipova S.I., Metel'skaya M.A. 1996. Ilnesses of children suffering from the Chernobyl catastrophe, 1994. Medical-biological consequences of the Chernobyl accident. № 1. pp. 38 – 47 (in Russian).

Lukomsky I.V., Protas R.N., Alexeenko Yu.V. 1993. Peculiarity of the neurological diseases of the adult population in the zone of the tight radiation control. "Impact of radionuclides' pollution on public health: clinical-experimental study". Collection of Sci. papers, Vitebsk State Med. Institute , Vitebsk, pp. 90—92 (in Russian).

Lukianova A.G., Lenskaya R.V. 1996. Dynamics of cytological-chemical characters of lymphocytes from the peripheral blood at Chernobyl children, 1987 -1995. Hematology and Transfusiology. vol. 41, № 6, pp. 27 – 30 (in Russian).

Malko M.V. 2002. Chernobyl Radiation-induced Thyroid Cancers in Belarus. In: Imanaka T. (Ed.). Recent Research Acticities about the Chernobyl NPP Accident in Belarus, Ukraine and Russia. Kioto Univ. Res. Reactor Inst. (KURRI-KR-79), Kyoto, pp. 240- 256.

Malko M.V. 2004. Radiogenic thyroid cancer in Belarus as consequences of the Chernobyl accident. In: Med.-biol. Problems radio- and chemical protection. Collect. of Papers Russ. Sci. conf., Sank-Petersburg, 20 – 21 May, 2004. pp. 113-114 (in Russian).

Manak N.A., Rusetskaya V.G., Lazjuk D.G. 1996. Analysis of blood circulatory illnesses of Belarus population .Medical-biological aspects of the Chernobyl accident. № 1, pp. 24 – 29 (in Russian).

Marples D.R. 1996. The decade of despair. Bull. Atomic Sci., May/June, pp.22—31.

Matveev V.A., Voropaev E.V., Kolomiets N.D. 1995. Role of the herpes virus infections in the infant mortality of Gomel territories with different density of radionuclides pollution. Actual problems of immunology and allergology. III Congress Belarus. Sci. Soc. Immun., Allergol. Grodno, Abstracts, p. 90 (in Russian).

Maznik N.A., Vinnikov V.A. 2002. Level of chromosomal aberrations in peripheral blood lymphocytes at evacuees and living in the radioactively polluted territories after the Chernobyl accident. Rad. Biol., Radioecol., vol. 42, № 6, pp. 704 – 710 (in Russian).

Mel'nichenko E.M., Cheshko N.N. 1997. Condition of children tooth and stomathological support in the regions of radioactive pollution. Public Health, № 5, pp. 38 – 40 (in Russian).

Mel'nov S.B., Senerichina S.E., Savitsky V.P., Dudarenko O.I. 1999. Medical- genetical aspects of the thyroid cancer at children after the Chernobyl accident. In: Ecological Anthropology. Almanac. Minsk, Belarus Committee "Chernobyl Children" , pp. 293 – 297 (in Russian).

Mezhzherin V.A. 1996. Civilization and Noosphera. Book 1. Kiev, 144 p. (in Russain).

Miksha Ya.S,, Danylov I.P. 1997. Consequences of the chronic impact by ionizing irradiation on the haemopoiesis in Gomel area. Public Health, № 4. pp. 19 – 20 (in Russian).

Morozevich T.S., Gres' N.A., Arynchyn A.N., Petrova V.S. 1997. Some eco-pathogenic problems of disturbances of the hair growth at Byelorussian children. Actual problems of medical rehabilitation of population suffering from the Chernobyl catastrophe. Materials Sci.- Pract. Conf. devoted by 10 anniversary Republic Radiation medicine dispenser, 30 June 1997, Minsk. Pp. 38 – 39 (in Russian).

Moskalenko B. 2003. Estimation of consequences of the Chernobyl catastrophe for Ukrainian public health. The World Ecological Herald, vol. XIV, № 3 - 4, pp. 4 – 7 (in Russian).

Nagornaya A.M. 1995. Adult population health of Zhytomir area, which suffer from radioactive impact after the Chernobyl accident and living in the strict control radiation zone (by National register data). «Public health problems and perspectives of Zhytomir area», Materials Sci.- Pract. Conf. devoted 100-years anniversary O.F. Gerbachevsky' hospital, Zhytomir, 14 September, 1995", Zhytomir, pp. 58 – 60 (in Ukrainian).

Nedvechkaya V.V., Lialykov S.A. 1994. Cardio-interval-graphical investigation of the nervous system at children from the radioactively polluted territories. Belarusian Public Health, № 2, pp. 30 – 33 (in Russian).

Nesterenko V.B., Yakovlev V.A., Nazarov A.G. (Eds.). 1993. Chernobyl catastrophe. Reasons and consequencies (Expert conclusion). Part 4. Consequences for Ukraine and Russia. Minsk, "Test" Publ., 243 p. (in Russian).

Nesterenko V.B. 1996. Scale and consequences of the Chernobyl catastrophe for Belarus, Ukraine and Rassia. Minsk, «The Right and Economics» Publ., 72 p.(in Russian).

Nyagu A.I. 1995. Nervous system. In: Barhyar V.G. . (Ed.). Chernobyl catastrophe. Kiev, "Naukova Dumka" Publ., pp. 458 - 459 (in Russian).

Nyagu A.I., Loganovsky K.N. 1998. Neuro-psychiatric effects of ionizing irradiation. Kiev, Scientific Center of Radiation Medicine Ukrainian Academy of Medical Science, 370 p. (in Russian).

Nyagu A.I., Loganovsky K.N., Loganovskaya T.K., Repin V.S., Nechaev S.Yu. 2002. Intelligence and Brain Damage in Children Acutely Irradiated in Utero As a Result of the Chernobyl Accident. In: Imanaka T. (Ed.). Recent Research Activities about the Chernobyl NPP Accident in Belarus, Ukraine and Russia. Kyoto Univ. Res. Reactor Inst. (KURRI-KR-79), Kyoto, pp. 202 – 231.

Okeanov A. E., Yakimovich G. V., Zolotko N. I., Kulinkina V. V. 1996. Dynamics of malignant neoplasms incidence in Belarus, 1974-1995. Biomedical aspects of Chernobyl NPP accident. No. 1, pp. 4 - 14 (in Russian).

Okeanov N.N., Yakimovich A.V. 1999. Incidence of malignant neoplasms in population of Gomel Region following the Chernobyl accident. Int. J. of Radiat. Medicine, vol. 1, No 1, pp. 49 –54 (cit. by R.I. Goncharova, 2000).

Okeanov A.E., Sosnovskaya E.Y., Priatkina O.P. 2004. A national center registry to asses trends after the Chernobyl accident. Swiss Med. Weekly, # 134, pp. 645 – 649.

Omelianetz N.I., Kartashova S.S., Dubovaya N.F., Savchenko A.B. 2001. Cancer mortality and its impact on life expectancy in the radioactively polluted territories of Ukraine. 3rd Inter. Conf. "Medical consequences of the Chernobyl catastrophe: results of the 15-years investigations. 4 - 8 June 2001, Kiev, Ukraine". Abstracts, Kiev, pp. 254 – 255 (in Russian).

Omelianetz N.I., Klement'eva A.A. 2001. Mortality and longevity analysis of Ukrainian population after the Chernobyl catastrophe. 3rd Inter. Conf. "Medical consequences of the Chernobyl catastrophe: results of the 15-years investigations. 4 - 8 June 2001, Kiev, Ukraine". Abstracts, Kiev, pp. 255 – 256 (in Russian).

Orlov Yu.A. 1993. Dynamics of congenital malformations and primitive neuro-ectodermal tumors. Social-psychological and psycho-neurological consequences of the Chernobyl catastrophe. Materials Sci. Conf. CIS states with inter. Participation, Kiev, pp. 259 -260 (in Russian).

Orlov Yu.A. 1995. Neuro-surgery pathologies' structure at children in post-Chernobyl period. Materials Inter, Sci. Conf. "Actual and prognostic infringements of psychic health after the nuclear catastrophe in Chernobyl, 24 -28 May, 1995. Kiev, Ukraine", Kiev, "Chernobyl Doctors" Assoc., p. 298 (in Russian).

Orlov Yu.A., Verchogliadova T.L., Plavsky N.V., Malysheva T.A., Shaversky A.V., Guslitzer L.N. 2001. CNS tumors at children (Ukrainian morbidity for 25 years). 3rd Inter. Conf. "Medical consequences of the Chernobyl catastrophe: results of the 15-years investigations. 4 - 8 June 2001, Kiev, Ukraine". Abstracts, Kiev, pp. 258 (in Russian).

Paramey V.T., Saley M.Ya., Madekin A.S., Otlivanchik I.A. 1993. Lens condition at people living in the radionuclides polluted territories.

Pasechnik L.I., Chuprykov A.G. 1993. Impact of radiation factor into formation of the children' neuro-psychical sphere. In: "Chernobyl catastrophe: diagnostics and medical-psychological rehabilitation of suffers", Collection of Conf. Materials, Minsk, pp. 15—16 (in Russian).

Paramonova N.S., Nedvetskaya V.V. 1993. Physical and sexual development indices of children under long-term impact of low doses irradiation. In: "Chernobyl catastrophe: diagnostics and medical-psychological rehabilitation of suffers", Collection of Conf. Materials, Minsk, pp. 62—64 (in Russian).

Pelevina I.I., Afanasiev G.G., Gotlib V.Ya., Serebryanyi A.M. 1996. Cytogenetical changes in peripheral blood at people living in the Chernobyl polluted areas. In: E.B. Burlakova (Ed.). Consequencies of the Chernobyl catastrophe. Public health. Moscow, Center for Russian Ecological Policy, pp. 229 – 244 (in Russian).

Petrenko S.V., Zaitzev V.A., Balakleevskaya V.G. 1993. Hypóphysis- adrenal system at children living in the radionuclides' polluted territories. Belarusian Public Health, № 11, pp. 7 - 9 (in Russian).

Petrova A.M., Maistrova I.N., Zafranskaya M.M. 1993. Infant' immune system in the territories with different levels of Cs-137 soil pollution. In: "Chernobyl catastrophe: diagnostics and medical-psychological rehabilitation of suffers", Collection of Conf. Materials, Minsk, pp. 74—76 (in Russian).

Pilinska M.A. 1999. Cytogenetic effects in somatic sells of people have suffered from the Chernobyl catastrophe as biomarkers of the low dose ionizing radiation impact. Inter. J. Rad. Med. , № 2, pp. 60 – 66 (in Russian).

Pilinska M.A., Dyb'skyi S.S., Dyb'ska O.B., Pedan L.R. 2003. Somatic chromosomal mutagenesis at children living in the radionuclides' polluted territories of Ukraine during post-Chernobyl period. Report Nat. Sci. Acad. Ukraine, № 7, pp. 176 – 182 (in Ukrainian).

Podpalov V.P. 1994. Formation hypertonic disease in population of the radioactively unsafe territories. In: "Chernobyl catastrophe: diagnostics and medical-psychological rehabilitation of suffers", Collection of Conf. Materials, Minsk, pp. 27—28 (in Russian).

Polonetskaya S.N., Chakolva N.N., Demedchik Yu.E., Michalevich L.S. 2001. Cytogenetical analysis of the normal and tumor thyroid gland cells in vivo. 4th Congress on Radiation Researches. Rad. Biol. Radioecol., Rad. Safety. Moscow, 20 – 24 November 2001, Abstracts, vol.1, p. 257 (in Russian).

Prilipko L.L., Nyagu A.I., Kozlova I.A., Gaiduk F.M., Loganovsky K.N., Podcorytov V.S., Plachinda Yu.I., Antipchuk E.Yu. 1995. Scientific resultes of researches on pilot project progrmme IFECA «in utero brain' infringements». Materials Inter. Sci. Conf. "Actual and prognostic infringements of psychic health after the nuclear catastrophe in Chernobyl, 24 -28 May, 1995. Kiev, Ukraine", Kiev, "Chernobyl Doctors" Assoc., p. 316 (in Russian).

Prokopenko N.A. 2003. Cardio-vascular and nervous systems pathologies as synergic result of the irradiation and psycho-emotional stress at sufferings from the Chernobyl accident. Aging and longevity problems. vol. 12, № 2, pp. 213 – 218 (in Russian).

Prysyazhnyuk A.Ye., Grishtshenko V.G., Fedorenko Z.P., Gulak L.O., Fuzik M.M. 2002. Review of Epidemiological Finding in Study of Medical Consequences of the Chernobyl Accident in Ukrainian Population In: Imanaka T. (Ed.). Recent Research Activities about the Chernobyl NPP Accadent in Belarus, Ukraine and Russia. Kyoto Univ. Res. Reactor Inst. (KURRI-KR-79), Kyoto, pp. 188 – 287.

Pshenichnikov B.V. 1996. Low doses radioactive irradiation and radiation sclerosis. Kiev, «Soborna Ukraina» Publ., 40 p. (in Russian).

Pysarenko S.S. 2003. On man sterility in XX Century. Herald New Med. Technology., vol. 10, № 3, pp. 106 – 107 (in Russian).

Romanenko A., Lee C., Yamamoto S. et al. 1999. Urinary bladder lesions after the Chernobyl accident: immune-histochemical assessment of proliferating cell nuclear antigen, cyclin D1 and P 21 wafl/Cip. Japan J. Cancer Res., vol. 90, pp. 144 - 153.

Rudnytzkyi E.A., Sobolev A.V., Kiseleva L.F. 2003. People illnesses by microsporidias in the radionuclides' zone. Problems Med. Micol., vol. 5, № 2, pp. 68 (in Russian).

Sevan'kaev A.V., Zhloba A.A., Potetnya O.I., Anykyna M.A., Moiseenko V.V. 1995. Cytogenetic observations at children and adolescens living in the radionuclides polluted territories of Bryansk area. Rad. Biol., Radioecol., vol 35, # 5, pp. 596 – 611 (in Russian).

Sevbitov A.V., Pankratiova N.V., Slabkovskaya A.B., Scatova E.A. 1999. Tooth – jaw' anomalies at children after impact of the "Chernobyl factor". Ecological anthropology. Almanac. Minsk, Belarusian committee «Chernobyl children», pp. 188 -191 (in Russian).

Sevbitov A.V. 2005. Stomatological characters of the clinical manifestations of the remote effects of irradiation. Thesis, Doc. Med. Sci., Central Stomatological Inst., Moscow, 51 p. (in Russian).

Sergeeva M.E., Muratova N.A., Bondarenko G.N. 2005. Demographic peculiarities in the radioactively polluted zone of Bryansk area. Materials Inter. Sci. – Pract. Conf.: «Chernobyl - 20 years after. Socio-economical problems and perspectives for development of suffering territories», Bryansk, pp. 302 – 304 (in Russian).

Sharapov A.N. 2001. Regulations of the endocrine – neuro-vegetative' interconnections at children living in the low doses radionuclides pollution territories after the Chernobyl accident. Thesis, Doc. Med., Sci. Inst.Pediat., Child. Surgery, Moscow, 53 p. (in Russian).

Shilko A.N., Taptunova A.I., Iskritzkyi A.M., Tschadystov A.G. 1993. Frequencies and etiology sterility and невынашивания in the Chernobyl factors impacted territories. Conf.: "Chernobyl catastrophe: diagnostics and medical-psychological rehabilitation of suffers.", Collection of Materials, Minsk, p. 65 (in Russian).

Sitnikov V.P., Kunitskyi V.S., Bakanova V.A. 1993. Clinical-immunological expressions of the oto-pharynx deceases at children from the Chernobyl zone. Radionuclides pollution impact to public health clinical-experimental study). Collection of Sci. Papers , Vitebsk State Med. Inst., Vitebsk, pp. 127—130 (in Russian).

Sluchik V.M., Kovalchuk L.E., Bratyvnych L.I., Shutak V.I. 2001. Cytogenetic effects of the low dose of ionizing radiation and chemical factors (15 years after the Chernobyl). 3rd Inter. Conf. " Med. Consequences of the Chernobyl catastrophe: results 15-years researches. 4 - 8 June 2001, Kiev, Ukraine", Abstracts, Kiev, p. 290 (in Russian).

Sokolov V.V. 2003. Retrospect estimation irradiated doses at the Chernobyl radioactively polluted territories. Thesis, Doc. Thechn. Sci. , Tula State University, Tula, 36 p. (in Russian).

Sokolovskaya Ya. 1997. One more Chernobyl impact. Radiation Pstruck do not only heart and blood, but brain. «Izvestia», 3 October, p.5 (in Russian).

Soloshenko EN. 2002. Immune homeostasis at the dermatitis patients, which suffers from radioactive irradiation during the Chernobyl accident. Ukrainian J. Hematol. Transfusiol., № 5, pp. 34 – 35 (in Ukranian).

Souchkevitch G. 1996. WHO tracks health effects. Monitor, vol. 4, №1, pp. 4—5.

Sources and effects of Ionizing radiation. 2000. UNSCEAR 2000 Report to the General Assembly. Annex J. Exposure and effects of the Chernobyl accident. United Nations, New York, 155 p.

Sources and effects of Ionizing radiation. 2000. UNSCEAR 2000 Report to the General Assembly. Annex G. United Nations, New York, 130 p.

State of Health at Bryansk population suffering from the Chernobyl accident. 1999. Collection analytical and statistical materials, 1995 - 1998. Bryansk, Department of Public Health, Bryansk Administration (http://www.miac.brk.ru).

Stepanov A.V. 1993. Analysis of the Trichocephalisis occurrence in the radioactively polluted territories. Radionuclides pollution' impact in public health (clinical-experimental study). Collection of Sci. papers, Vutebsk State Med. Inst., Vitebsk, pp. 120 – 124 (in Russian).

Stepanova E.I., Davidenko O.A. 1995. Children haemopoetic system' reactions on the unhealthy Chernobyl accident' impact. III Ukr. Congr. Hematol. Transfusiol. 23-25 May, 1995, Sumy, Ukraine", Kiev, p. 134 (in Ukrainian).

Stepanova E.I, Misharina Zh.A., Vdovenko V.Yu. 2002. Remote cytogenetics effects at in utero irradiated children after the Chernobyl accident. Rad. Biol., Radioecol., vol. 42, № 6, pp. 700 – 703 (in Russian).

Strukov E.L. 2003. Hormonal regulation of the cardio-circular diseases and some endocrine dysfunctions at persons suffering by the Chernobyl factors and in Sank-Petersburg population. Thesis Doc. Med., All-Russian Center for Extr. Rad. Medicine, Sank-Petersburg, 42 p. (in Russian).

Surykov B.T. 1996. Chernobyl: 10 years the largest technological catastrophe in history. Ecology and Life, № 1. pp. 31—36 (in Russian).

Sykorenskyi A.V., Bagel' G.E. 1992. Primordial arterial hypotonic at children of Gomel and Mogiliev areas and perspectives of their improvement in the summer camps. Improvement and sanitary treatment of persons suffering from radiation impact. Abstracts, Republican Conf., Minsk - Gomel, pp. 59 – 60 (in Russian).

Tereschenko V.M., Geetz V.I., Perevosnikov O.N, Litvinetz L.A. 1991. Age characteristics of the Cesium level at inhabitants of Narodychesky district of Zhytomir area. Probl. Rad. Medicine, Kiev, pp. 99 – 103 (in Russian).

Terletzkaya R.N. 2003. Lung' chronic deseases at children after long-time living under law dose impact. Rus. Herald Perinatol.,Pediatry, vol. 48, № 4, pp. 22 – 28 (in Russian).

Tron'ko N.D., Tchaban A.K., Oleinik V.A., Epstein E.V. 1995. Endocrine system. In: Baryachtar V.G. (Ed.). Chernobyl catastrophe. Kiev, "Naukova Dumka" Publ., pp. 454 – 456 (In Russian).

Tsybul'skaya I.S., Suhanhova L.P., Starostin B.M., Mitryukova L.B. 1992. Functional condition of the cardio-vascular system at children early ages under the low dose irradiation chronic impact. Motherhood and Childhood, vol. 37, № 12, pp. 12 – 20 (in Russian).

Ulyanova O.S., Mashneva N.I., Ponomarev A.V., Sukal'skaya S.Ya. 1995. Psychomotoric development of children, with different impact in utero Chernobyl irradiation. Inter. Conf. "Actual and prognostic infringements of psychic health after nuclear catastrophe in Chernobyl, 24 - 28 May, 1995, Kiev, Ukraine", Materials, Kiev, "Chernobyl Doctors" Assoc., p. 318 (in Russian).

Utka V.G., Scorkina E.V., CSadretdinova L.Sh. 2005. Medical-demographic dynamics in South-Western districts of Bryansk area. Materials Inter. Sci. – Pract. Conf. «Chernobyl – 20 years after. Socio-economical problems and perspective for development suffering territories», Bryansk, pp. 201 – 203 (in Russian).

Ushakov I.B., Arlaschenko N.I., Dolzhanov A.Ya., Popov V.I. 1997. Chernobyl: radiation psychophysiology and human ecology. State Sci. Inst. Avia, Space Med., Moscow, 247 p. (in Russian).

Ushakov I.B., Karpov V.N. 1997. Brain and radiation (100 anniversary of radio-neurobiology). State Inst. Avia , Space Medicine , Moscow, 74 p. (in Russian).

Ushakova T.N., Aksel' E.M., Bugaeva A.R., Maikova S.A., Durnoe L.A., Poliakov V.G., Symonov A.F. 2000. Peculiarities of the cancer tumors ilnewsses at children of Tula area after the Chernobyl catastrophe. In: «Chernobyl: Duty and courage», vol. I (http://www.iss.niiit.ru/book-4/glav-2-26.htm) (in Russian).

Vasyna T.I., Zubova T.N., Tarasova T.G. 2005. Some hematological characters at children, living in the territories polluted after Chernobyl accident. Inter. Sci.- pract. Conf.

«Chernobyl - 20 years after. Social-economical problems and perspectives for development of suffering territories», Bryansk, pp. 152 – 154 (in Russian).

Vaskevich A.Yu., Chernyshova V.I. 1994. Mozyr city' children health under low intensive radioactive pollution. Belarusian Children health in the modern ecological conditions (Chernobyl consequences). Collection of materials, V1 pediatric Congress of Belarus. Gomel, pp. 27 - 29 (in Russian).

Voropaev E.V., Matveev V.A., Zhavoronok S.V., Naralenkov V.A. 1996. Activation of Herpesvirus infection after the Chernobyl accident. Sci. conf. "10 years after the CChernobyl catastrophe: scientific aspects", Abstracts, Minsk, p. 65 (in Russian).

Vovk U.B., Mysurgyna O.A. 1994. Estimation radioactive pollution and doses of irradiation from the Chernobyl accident in the Global scale. Inter. Conf. "Nuclear Accidents and Future of Energetic. Chernobyl Lessons (Paris, France, 15 – 17 April, 1991)", Selected Papers, Minsk, pp. 120 - 144.

Yamashita S., Shibata Y., (Eds.). 1997. Chernobyl: A Decade. Proc. Fifth Chernobyl Sasakava Med. Coop. Symp., Kiev, 14 - 15 October, 1996, Elsevier Sci. Publ., Amsterdam.

Zalutskaya A., Bornstein S.R, Mokhort T., Garmaev D. 2004. Did the Chernobyl incident cause an increase in Type 1 diabetes mellitus incidence in children and adolescents? Diabetologia, vol. 47, pp. 147-148.

Zhavoronok S.V., Kalinin A.L., Grimbaum O.A., Chernovetskyi M.A., Babarykyna N.Z., Ospovat M.A. 1998. Gepatoviruses B, C, D, G markers at suffering from the Chernobyl catastrophe populations. Public Health, № 8, pp. 46 – 48 (in Russian).

Zakrevsky A.A., Nikulina L.I., Martynenko L.G. 1993. Early postnatal adaptation of newborns whose mothers were under radiation impact. Sci.- Pract. Conf.: "Chernobyl and Public Health". Kiev, Abstracts, Part 1, p. 116 (in Russian).

Zapesochnyi A.Z., Burdyga G.G., Tsybenko M.V. 1995. Irradiation in utero and intellectual development: complex science-metrical analysis of information flows. Materials Inter. Conf. "Actual and prognostic infrigments of psychical health after nuclear catastrophe in Chernobyl. 24 – 28 May 1995. Kiev, Ukraine", Kiev, p. 312 (in Russian).

Zubovich V.K., Petrov G.A, Beresten' S.A., Kil'chevskaya E.V., Zemskov V.N. 1998. Human milk characters and babies health in the radioactively polluted areas of Belarus. Public Health, № 5, pp. 28 – 30 (in Russian).

Yablokov A.V. 1997. Nuclear mythology. Ecologist' note on Nuclear Industry. Moscow, "Nauka" Publ., 272 p. (in Russian).

Yablokov A.V. 2001. Myth on insignificance of the Chernobyl catastrophe consequencies. Moscow< Center for Russ. Ecol. Policy Publ., 112 p. (in Russian).

CHAPTER 2

Is it Safe to Live in Territories Contaminated with Radioactivity
Consequences of the Chernobyl accident 20 years later

E.B. Burlakova[1] and A.G Nazarov[2]
[1]*Emanuel Institute of Biochemical Physics, Russian Academy of Sciences , Moscow*
[2] *"Union of Chernobyl" Moscow committee*
(Translated by T.A. Sapego)

Introduction

The key criterion for determining the severity of any catastrophe, natural or technogenic, is its impact on the health of people and the conditions of their continued habitation in the effected territories. The Chernobyl catastrophe, which is the severest one in human history, has been treated ambiguously. There are known efforts of the International Atomic Energy Agency (IAEA) and the Ministry of Atomic Energy of Russian Federation (Minatom) to underrate the consequences of the Chernobyl NPS accident on the health of both the liquidators (clean-up workers of the Chernobyl accident) and also residents of the affected regions. As is shown in this book (see Chapter 6), the IAEA experts, supported actively by the officials from the Minatom and USSR Ministry of Health and appealing to the results of their "independent expertise" (1988), declared that the territories within the zone of the Chernobyl accident are adequate for safe living. Similar claims are heard in statements made by IAEA officials now, 20 years after the accident. We do know the attitude of residents of the affected regions to such an assessment of conditions of "safe" living in the Chernobyl zone; In 1988-1989, public movements and social organizations formed and demanded the disclosure of secret information concerning the Chernobyl accident, the implementation of measures for the decontamination of the affected areas and the rendering of state help for the people who had suffered.

Studies of radiation effects on the health of the liquidators and the population of the Chernobyl NPS zone were initiated by several medico-biological institutions almost immediately after the accident. One of the first scientists who began stationary studies on the genetic consequences of the catastrophe was V.A. Shevchenko, an eminent Russian radiation geneticist, deceased from cancer not long ago.

During the first several years after the Chernobyl accident many aspects of radiation effects on the human organism remained unclear. The studies required a great number of laboratory experiments on animals; experiments involving various aspects of radiation effects were performed at many research institutes of the Academy of Sciences and other institutions. Between 1987-1998, at the Emanuel Institute of Biochemical Physics, Russian Academy of Sciences, a series of studies on the effect of low-dose low-level irradiation on biophysical and biochemical parameters of the genetic and membrane apparatus of the cells of organs of exposed animals were carried out.

At present, the experimental data obtained are being analyzed and generalized. We apologize to the reader for the abundance of scientific terms in this work that may only be familiar to specialists or radiobiologists. Later on we intend to convert the conclusions and experimental results, which, we believe, are of theoretical importance, to a language more easily understood by non-specialists. Some of the purely scientific concepts; methodological radiobiological argumentation and tabular data may be omitted by a non-specialist upon first reading. Nevertheless, we found it necessary to retain them in this book as new theoretical material – facts, accumulated for more than 10 years of experimental studies, giving evidence and confirmation of the inferences and conclusions made.

Analysis and Generalization of Experimental Data

We investigated the structural parameters of the genome (by the method of DNA binding to nitrocellulose filters), structural parameters of nuclear, microsomal, mitochondrial and plasmic (synaptic and erythrocyte) membranes (by the method of spin probes localized in various layers of membranes), the composition and oxidation degree of membrane lipids, and the functional activity of cells – the activity of enzymes, relationships between isozymic forms and regulating properties [1, 2, 3].

We also investigated the effect of low-level irradiation on the sensitivity of cells and biopolymers to subsequent action of various damaging factors, including high-dose irradiation. The animals were exposed to a source of ^{137} Cs γ-radiation at the dose-rates 41.6 x 10^{-3}, 4.16 x 10^{-3}, and 0.416 x 10^{-3} mGy. The doses were varied from 6 x 10^{-4} to 1.2 Gy.

As a result of the studies performed, the following conclusions were made:

1. The dose dependence of the radiation effect may be non-linear, non-monotonic, and polymodal in character.
2. Doses that cause extreme effects depend on the irradiation dose-rate (intensity).
3. Low-dose irradiation causes changes (mainly enhancement) in sensitivity to the action of other damaging factors.
4. The effects depend on the initial parameters of biological objects.
5. Over certain dose ranges, low-level irradiation is more effective with regard to the results of its action on an organism or a population than acute high-level radiation.

We explain the non-linear and non-monotonic dose–effect dependence that we obtained in our experiments with low-dose low-level irradiation by changes in the relationship between damages, on the one hand, and reparation of the damages, on the other hand [4]. With this kind of irradiation the reparative systems either are not initiated (induced), or function inadequately, or are initiated with a delay, i.e., when the exposed object has already received radiation damages.

It is difficult to predict the dose dependence for the effect, which is a result of the interaction of several subsystems, when each of the subsystems is sensitive to a certain factor and exhibits its own characteristic response to increasing dose.

Generally speaking, one can hardly expect a monotonic increase in the resulting radiation effect with increasing dose because the determining factor here is not solely the reaction of each individual subsystem but the sign (direction) and character of their interaction.

In particular, the experiments showed that the radiation effect on an organism, along with its direct action on the structural and functional biological subsystems, mobilizes and activates the protective systems of reparation, adaptation, etc., whose regulating role is compensation and minimization of the direct irradiation effect, restoration of functions, and repairing the damages. After the initiation of reparative processes, the resulting (residual) effect depends on the relationship between the direct (irradiation) and reverse (restoration and compensation) processes, which is different for each separate irradiation dose.

Recently, the absence of reparation (the mechanism of repair of damages) at low irradiation doses was shown at the cell level, and the complex character of the dose dependence was confirmed [5, 6]. Previously, we published a scheme of dependence of damages on irradiation dose, which shows the differences for different dose ranges (see [1, 2, 4]). According to the scheme, the quantitative characteristics were similar for the doses that differed by several orders of magnitude; in a certain dose range the effect was opposite.

The results obtained and supported by numerous experiments are important because the above dose dependences make it possible to come to a conclusion about the radiogenic or non-radiogenic character of changes observed in an irradiated organism. Along with the overwhelming majority of researchers, we think it indisputable to conclude that a monotonic, nearly linear, increase in the irradiation effect on an organism with increasing irradiation dose is (or, at the present level of knowledge about high dose effects, may be) evidence for its radiogenic nature.

Nevertheless, the results of many years of our experimental studies do not favour the opposite conclusion: that the absence of a direct dose dependence and its non monotonic character is evidence for the absence of the relation between the effect and radiation. In our opinion this relation and the radiogenic nature of the effect are equally indisputable, as is the case with high doses: it is mainly that this character of the dose–effect dependence seems natural in the case of low-dose low-level irradiation.

Often, for processes that play a certain role in damaging biomacromolecules, one should consider their participation, not only in damaging or restoring events, but also their role in the regulating network of cells that determine the character of response to irradiation. For example; active oxygen species may participate in normal metabolic processes, radiation damage (of DNA molecules, proteins, lipids, and membranes) and regulating processes in cells responsible for division, reproduction (proliferation), differentiation and programmed death of cells (apoptosis).

It is difficult to predict the resulting overall response of a cell without the knowledge of the response of each of its systems and their dependence on the intensity, mode and dose of irradiation and the overall nature of their radio-sensitivity. However, it is possible to make a general conclusion that the resulting response of an organism, cell or population will depend, to a large extent, on the balance of opposite processes, e.g., apoptosis and proliferation.

In the article of Hardy and Start [7], a mathematical model of such a balance is suggested; the model allows for the behavior of a cell after irradiation: the cell will be either:

- Ruined, with the probability **P**(apoptosis), or
- Divided, with the probability **P**(proliferation)
- Or will remain unchanged, with the probability **P**(no change).
- The balance of these processes will determine the character of the dose dependence.

Experimentally, we studied the trends of changes in three parameters related to extreme points on the dose dependence curves:

- Time, during which the effect reaches the maximum;
- Dose, at which the maximum is reached;

-and the effect at the extreme points, depending on the irradiation intensity, varied by an order to two orders of magnitude (from 0.06 to 0.6 and 6.0 cGy).

From rates of change in the DNA structural parameters we determined that a decrease in the irradiation intensity results in an increase in the time of the effect reaching its maximum, a decrease in the dose of the maximum effect, and a decrease in the effect at the extreme point. Similar trends were determined for nearly all parameters of membranes. Exceptions were the data obtained for the lowest irradiation intensity, at which maximum effects were observed. With regard to the fact that the maximum effects are observed at the time of initiation of the reparative systems, the trends we determined make it possible to conclude that the lower the irradiation intensity, the later the reparative systems are initiated.

The results obtained are evidence for a high biological activity of low dose irradiation and different mechanisms on the cell metabolism than those for high doses.

Another theoretical generalization of the experimental data, which is of great practical importance, is the phenomenon that we and other authors noted: enhancement of the sensitivity of living organisms irradiated with low doses and enhancement of the susceptibility of biochemical processes to the subsequent action of damaging agents. This phenomenon may be explained plausibly by the radiation-induced instability of the genome.

Note that the genome restructuring and, as a result changes in the accessibility of the genetic material to regulating effects, play an important role in these processes. In our work [8] and the work of others [9, 10], the expression of regulating genes (initiation, induction, or starting functioning) after low-dose low-level irradiation of organisms was observed. These findings are of great importance because changes in the sensitivity to the action of many other damaging factors after low dose exposures may be (and actually are) the cause of diseases and disturbances in the adaptive ability of man. It should be emphasized that these processes are closely related to aging, which is a process also characterized by enhancement of the sensitivity to damaging factors and the probability of death from these factors increases with age.

We do not intend to give a full explanation regarding changes in biochemical and biophysical processes occurring in organisms exposed to low doses. However, the data presented here show that these changes may cause various somatic diseases. To determine the radiogenic nature of these diseases it is not necessary to perform mathematical processing of the dose dependences over the entire range of doses received. The criteria for a conclusion regarding the radiogenic nature of diseases should be worked out by taking into account specific radiation-induced disturbances in the cell metabolism at the cell, population, organ, and organism levels.

It is important to study the effect of low-level irradiation at the population level. The primary effect detectable in various systems is a change (increase) in the variance in many population indices and decrease in stability, including adaptation. For example; in an exposed human population the number of people whose blood cells do not give an adaptive response to subsequent exposures has increased [11]. The correlation between indices, which is most pronounced for groups who received low doses of radiation, varies. Our studies showed that the correlations between indices of the antioxidant and immune status of ChNPS liquidators were changed [1].

Radiation-induced changes in the population structure result in an unpredictable response of the population to various events. In the work by A.P. Akif'ev et al. [12], an apparently healthy population of the posterity of exposed *Drosophila* exhibited a so-called 'populational breakdown' in one of its generations and was ruined by a law other than that for other generations. In the work by I.I. Pelevina et al. [13], it was shown that 15 generations of cells irradiated with the doses 10 to 50 cGy "remember" the irradiation and respond to external stimuli differently than the control.

According to A.A. Yarilin [14], low-level ionizing radiation is a source of biologically significant signals. Having analyzed changes in the immune system from this point of view, the author came to the conclusion that inadequate signals brought about by low-level irradiation result in disturbances in the spatial organization of the immune system and its integrating functions in the organism. In a sense, the effect is similar to the aging of the immune system and in the organism.

In this work, much attention is given to free radical reactions caused by irradiation, in particular promotion of oxidation of fats (lipids) and associated changes in the composition and functional activity of membranes, restructuring of the membrane apparatus, increase in the concentration of free radicals in various components of cells, antioxidant activity of regulating enzymes, changes in physicochemical properties and regulation of the activity of the genome (expression and repression of genes).

In experiments in animals and in studies on biochemical parameters of formal elements and blood plasma of man, common trends in the effects of low-dose low-level irradiation were observed, namely disturbances in the correlations between oxidizability and antioxidant properties of lipids and between structural changes in lipids localized in various portions of membranes [15]. These changes result in a loss of the regulating functions of membranes. Similar trends were observed for structural changes in DNA and the genome [16]. At present, there are many works published that verify the crucial role of the signaling functions of active oxygen species in the regulating network of cell response to damaging impacts, radio-sensitivity and instability of the genome.

Therefore, changes in the composition, structure, and functional activity of membranes are primary symptoms of disturbances in the cell metabolism and a predicting factor in the development of a disease.

The above trends follow from the generalization of a series of experiments and are of a common biological character. Therefore, they may be used in analyzing the state of health of Chernobyl residents and answering the question whether it is safe to live on the radiation-contaminated territories within the zone of the Chernobyl accident.

Table 1. Parameters of the antioxidant status for the liquidators before and after one month vitamin therapy

Parameter	Control	Liquidators before therapy	Liquidators after therapy
DBpl (double bonds in plasma lipids, number of DB/mg lipids 1018)	0.32	0.27	0.39
DBer (double bonds in erythrocyte lipids, number of DB/mg lipids 1018)	0.303	0.15	0.47
Vitamin E	20.9	15.59	21.59
Vitamin A	2.99	2.67	2.82
Reduced glutathione	19.53	20.34	15.32
SOD (superoxydismutase)	125.41	137.53	105.30
GP (glutathione peroxydase)	7.2	9.28	4.93
GR (glutathione reductase)	5.12	5.75	5.97
Hem1 (hemolysis of erythrocytes)	7.23	5.79	8.65
Hem2 (hemolysis of erythrocytes after initiation of POL)	7.62	11.32	11.04
MDA1 (malonic dialdehyde in erythrocytes)	1.93	3.90	1.89
MDA2 (malonic dialdehyde in erythrocytes after initiation of	1.95	2.58	1.89

POL)			
τcl (time of rotary correlation of spin probe N1 in erythrocyte membranes)	1.08	2.06	1.04
τclI (time of rotary correlation of spin probe N2 in erythrocyte membranes)	1.94	2.22	1.63
CP (ceruloplasmine)	1.23	0.80	0.86
TF (transferine)	0.78	1.03	0.77
Free radicals with g-factor 2.0	0.69	2.03	1.14

Table 1 shows some indices of the state of health of liquidators who worked in the zone of the CNPS accident in 1986-1987.

Along with parameters of the antioxidant (AO) status of the liquidators, we measured the parameters of the immune status, which also were found to be significantly below normal, even five years after the irradiation [3]. At different times after the accident, we investigated erythrocytes and blood plasma from 104 liquidators who worked in Chernobyl between 1986-1987 and received irradiation doses from 0.1 to 150 cSv. We established a decrease in the level of antioxidants (tocopherol, vitamin A, and ceruroplasmine), enhancement in the concentration of the products of free radical reactions and a high level of free radicals, a more rigid membrane, and a break in the correlation between oxidizability and AO activity and viscosity of the lipid and protein components.

Some of the liquidators were administered with antioxidants (vitamins) for a month and then were examined. It was shown that 70% of indices of the AO and immune status were normal after the antioxidant therapy.

It was important to examine liquidators of various ages to determine their reaction to the irradiation doses received. In fact, the content of antioxidants decreases and the activity of protective antioxidant enzymes varies with age. Six years after the accident we determined the age-related dependences of the activity of the key AO enzymes in the blood (superoxydismutase, glutathione peroxidase, and glutathione reductase) of liquidators between the ages of 25 to 60 years. The control was 35 men and women from the same age group.

For liquidators, enzymes of the glutathione cycle were the most sensitive to low dose radiation links with the AO system. Changes in all links of the AO system of liquidators promote formation of the pro-oxidant state. According to our data, doses above average produce a prolonged damaging effect on the AO system. As was noted above, a decrease in the SOD and GP activity in middle-aged liquidators and a drastic decrease in the GR activity in liquidators older than 55 years were recorded over the entire dose range studied. Previously, we showed that pre-cancer changes in the cell metabolism are characterized by a decrease in the SOD/GP relationship and a low level of GR activity. Both these indices were recorded for the older liquidators.

The results obtained in this work on the remote consequences of low-level low-dose irradiation on the protective AO system of people show that the most sensitive members of a population are children and young people below the age of 30 years; middle-aged people are the most resistant to irradiation. This conclusion should be taken into account when determining high risk groups for people engaged in industries involving chronic low-level irradiation.

As for young people, low doses of low-level irradiation cause an imbalance in the AO system, which is characteristic of an aging organism [17].

In the works by L.S. Baleva *et al.,* some of these parameters were measured for children living within the contaminated territories. An enhanced concentration of the products of peroxide oxidation of lipids (POL) and a lower activity of AO enzymes of the glutathione cycle were detected. The most drastic changes were detected in children born in 1986-1987 who remained living within the radionuclide-contaminated territories [18]; serious deviations from normal were detected also for children born before 1986 who survived the accident and remained living within the contaminated territories (Table 2).

Table 2 Biological monitoring of children of the Bryansk region (Baleva and Sipyagina [18])

Parameters	Groups of children			
	Born before July 4, 1986.	*In utero* irradiation 1986-1987	Intrauterine development under conditions of elevated radiation background 1988	Control
MDA, nmol/ml	4.55±0.04	4.96±0.05	4.49±0.07	3.56±0.01
Hydroperoxides, a.u./ml	2.06±0.03	1.89±0.05	1.81±0.06	1.5±0.02
AOA, %	30.86±0.14	29.34±0.18	33.1±0.21	34.2±0.28
GP	.69±0.07	2.81±0.12	3.38±0.18	3.94±0.20
SOD	121.9±8.8	61.4±6.3	114.0±9.2	85.2±5.1
CP, a.u.	53.1±5.7	69.8±3.3	112.3±6.5	86.1±8.7
Fe (3+), TF a.u.	72.8±6.5	78.9±5.1	55.2±5.6	56.0±4.8
α-1 antitrypsin, a.u./ml	11.72±0.45	3.89±0.39	8.91±0.51	23.71±0.31
Extracellular DNA in blood plasma (mg/ml)	5.91±0.37	7.87±0.58	3.08±0.36	4.51±0.25
Content of endonuclease in blood plasma, a.u.	36.8±1.1	55.2±1.3	49.1±1.5	42.0±1.2

The Medical Radiological Research Center, Russian Academy of Medical Sciences, performs constant monitoring on the state of health of liquidators and residents of the contaminated regions. Surveys (1991-1996) revealed a drastic aggravation of the health of liquidators during this period. In 1991, about 20% of liquidators were assigned to Group I (apparently healthy); 50%, to Group II; and 27%, to Group III. In 1996, 8% were apparently healthy (Group I) and 68% of people had three or more chronic diseases (Group III). A still more serious situation was recorded in 2002-2003. Among the liquidators living in Moscow and the Moscow region, none were apparently healthy; 100% had three or more chronic diseases. For St. Petersburg and Leningrad Region, 85% were assigned to Group III. The number of liquidator-invalids was 37% (31% in 1999); the Chernobyl-related invalidity accounted for 95%. The most frequent causes of invalidity were diseases of the central nervous system (70%), blood circulation system (23%), and locomotor apparatus (6%). Polymorbidity, i.e., three or more chronic diseases, was recorded for 59% [19].

A comparison of the data obtained in the Obninsk Medical Center in 1993 and 2003 on different diseases among liquidators and the population of Russia (per 100,000) showed that the above tendencies persist [20, 21], which is evident from Table 3.

Table 3 Liquidators to population of Russia morbidity ratio (1993 and 2003)

Classes of diseases	Ratio 1993 [20]	Ratio 2003 [21]
Neoplasms	0.9	-
Malignant diseases	1.6	-
Endocrine diseases	18.4	-
Diseases of blood and hemopoietic organs	3.6	4.55
Mental diseases	9.6	9.95
Diseases of blood circulation system	4.3	-
Gastrointestinal diseases	3.7	1.9
All classes of diseases	1.5	1.59

The groups of incipient cases among liquidators exceed those for the population of Russia by a factor of 5 to 10 with regard to diseases of the central nervous system (CNS) and cardiovascular diseases. The data on three Russian regions, where an analysis of chronic diseases was performed in 2001-2003 (Table 4), were kindly supplied by Dr. I.V. Oradovskaya.

Table 4 Incidence rate of chronic diseases of liquidators (monitoring data) in 2001-2003 [21]

Diseases		Moscow+ region		St.Petersb.+ region		Krasnoyarsk region	
		2003	2002	2003	2002	2003	2002
		n=110	n=133	n=104	n=108	n=74	n=194
Diseases of blood circulation system		98.18	85.71	85.58	72.22	85.14	81.44
1	Atherosclerosis+hypertensive disease	82.72	56.39	63.46	47.22	58.11	44.32
2	Coronary disease	71.81	48.87	40.38	43.52	36.49	26.80
3	VVD, NCD	13.64	25.56	14.42	13.89	21.62	25.77
4	CVD:Discirculatory encephalopathy	86.36	49.62	59.62	42.59	71.62	51.03
Pathologies of nervous and mental spheres		80.0	41.99	34.62	25.93	82.43	40.72
5	Asthenic syndrome, neurasthenia	53.63	20.30	15.38	9.26	13.51	9.28
6	Syndrome of high fatiguability	62.72	19.55	17.31	18.52	78.38	22.68
7	Organic diseases of craniocerebrum	14.54	12.78	5.77	2.78	25.68	14.43
8	Polyneuropathy	9.09	3.03	0.96	—	2.70	4.12
Diseases of alimentary organs		96.36	72.18	66.35	52.78	64.87	63.92
9	Diseases of gastrointestinal	95.45	51.88	56.73	47.22	47.30	42.27

	tract (chronic gastritis, gastroduodenitis, ulcer of stomach &duodenum)						
10	Chronic cholecystitis, cholecystopancreatitis	49.09	40.60	25.96	20.59	45.95	41.24
11	Fatty hepatosis & dystrophy of liver	17.27	5.26	9.62	5.56	16.22	8.69
Diseases of musculoskeletal system		100.0	66.17	53.85	53.7	52.70	52.06
12	Deforming osteochondrosis of backbone	91.81	56.39	48.08	46.30	48.65	44.32
13	Chronic polyarthritis, osteoarthrosis	39.09	14.29	1.92	6.48	8.11	7.22
14	Osteoarthrosis	31.82	12.78	10.58	6.48	1.35	1.56
Other chronic pathologies							
15	Diseases of veins	8.17	6.79	15.39	10.19	4.05	4.12
16	Multiple caries	10.0	6.02	0	3.72	1.35	4.12
17	Diseases of the thyroid	42.72	30.08	26.92	26.85	24.39	22.16
18	Non-infectious pathology of vision	50.0	9.77	8.65	8.33	16.22	6.19
19	Radiation-induced cataract	8.18	3.76	0	4.63	2.70	2.06
20	Non-infectious pathology of hearing	6.36	1.50	0.96	0.93	4.05	1.55
21	Non-allergic dishydrotic eczema	3.63	0.75	—	—	—	
22	Urolithiasis	26.36	16.03	10.58	11.11	13.51	7.22
23	Chronic pyelonephritis	4.54	—	—	—	1.35	0.52
24	Diabetes mellitus, type II	4.54	3.01	0	0.93	2.70	2.06
25	Benign tumors	35.45	14.10	12.50	16.67	13.51	4.12
26	States after resection of malignant tumors	2.72	1.50	3.85	2.78	4.05	4.12
27	Apparently healthy	0	0	0	0	0	2.06
28	Polymorbidity (3 and more diseases)	100.0	84.21	86.54	72.22	77.03	74.23
29	Chronic diseases (total)	100.0	96.99	100.0	89.81	86.49	97.94
30	Invalidity	35.45	33.08	26.92	25.93	43.24	38.66
	Including: Group I	0.9	0.75	0	0.93	4.05	1.55
	Group II	25.45	23.31	20.19	13.21	18.92	18.04
	Group III	9.09	8.27	5.77	12.04	20;27	19.07
31	General diseases	19.09	19.08	5.77	3.72	5.41	3.61
32	Total	54.55	51.13	32.69	29.63	48.65	42.27

It is evident from Table 4 that diseases specific for middle-aged people prevail. Indeed, the age of liquidators was several years younger than would be expected from the above assessment of their state of health.

In the works by A.F. Tsyb and V.K. Ivanov, *et al.* [22], dose dependences of non-cancer diseases of liquidators were considered.

The studies resulted in determining statistically significant dose dependences for the whole cohort of liquidators stratified by age at the time of arrival to the zone, date of arrival, and region of residence with regard to the following non-oncological classes of diseases:

- Diseases of the endocrine system [ERR = 0.58 with 95% CI (0.30 and 0.87)];
- Mental diseases [ERR = 0.40 with 95% CI (0.17 and 0.63)];
- Diseases of the central nervous system and sensory organs [ERR = 0.35 with 95% CI (0.19 and 0.52)];
- Diseases of alimentary organs [ERR = 0.24 with 95% CI (0.05 and 0.43)];
- Cerebrovascular diseases [ERR = 1.17 with 95% CI (0.45 and 1.88)];
- Essential hypertension [ERR = 0.52 with 95% CI (0.07 and 0.98)].

The results obtained should be considered as preliminary and requiring verification. Further surveys over the chosen cohort of liquidators will eliminate the ambiguity in the quantitative assessment of the results and will make it possible to isolate a radiation component in pathologies by taking into account all risk factors of both the radiogenic and non-radiogenic nature.

Foreign physicians are inclined to ascribe all the health troubles of the liquidators and the population (children and adults) to either faults in medical care and recording or adverse social conditions. It would be unjust to disregard the harshness of living conditions for the people of Russia. However, a local comparison of irradiated and non-irradiated contingents living under the same conditions and even engaged in harmful or heavy industries makes it possible to draw a conclusion about the obvious contribution of irradiation to the loss of health of the irradiated population, in particular children and liquidators. We have already noted that the relation of the effects with doses of low-level irradiation does not necessarily obey the same laws as those for high doses and that the effect may be not only non-linear but also non-monotonic. Hence, detecting the radiogenic nature of low and high-dose effects should be different; the same criteria and approaches do not hold. The criteria for determining the dose-effect correlation should be based on data of molecular epidemiology. At present, a promising approach is the search for relationships between somatic diseases and cytogenetic disturbances in the organism of irradiated people. The number of studies in which this relationship has been detected has grown rapidly.

Conclusion

A series of experimental medico-biological, biochemical, biophysical, and cytogenetic studies performed during the post-Chernobyl period with the use of data on the effect of the CNPS accident on the health of liquidators and the population of the radio-contaminated regions of Ukraine, Russia, and Belarus revealed two common biological trends. One of these shows with a high level of statistical confidence, the role and effect of low doses of low-level irradiation on man and living things. Another, closely related to the first one, shows an increase in the sensitivity of objects exposed to low level irradiation to other damaging factors including higher irradiation doses.

In examining these trends some others were discovered that show various specific aspects and phenomena of low-dose radiation impacts. In particular:

- The correlation between the destructive effects of irradiation and the restoration (repair) of damages.
- The decisive protecting role of cell membranes; the great importance of the AO stability and immune status toward the effects of low doses.
- The complicated nature of dose dependences on effects involving the interaction of several subsystems.
- The informative (signaling) nature of biologically significant low-level radiation stress.

- Specific features in the response of a population to the action of low doses.
- Others described in this work.

Although some of these findings require further studies, the trends and phenomena discovered may become a theoretical basis for making prognoses on the state of health of the victims of the Chernobyl accident and practical recommendations for improvement of the present situation.

The results of surveys and biological monitoring of children and adults of Chernobyl point unambiguously to a steady, rapid, and dramatic (for an individual human life) deterioration of health of all victims of the radiation impact of the Chernobyl accident. The effect manifests itself in the processes of rapid aging of the organism and the development of so-called polymorbidity, i.e., a complex of three or more chronic diseases, among victims of irradiation. At present, the extent of this pathology among liquidators living in Moscow and the Moscow region is 100%; for the Leningrad region, 85%.

An important result of the studies performed is the assessment of the effects of low doses of irradiation on people of different age categories. The most resistant to irradiation are middle-aged people - children and young people below the age of 30 and old people over 60 are the least resistant. This is an important practical finding; it makes it possible to work out prognoses for the rates of real employment for a population of various age groups in radio-contaminated regions and industries involving low-level irradiation.

The question put in the title - whether it is safe to live on radiation-contaminated territories - is controversial and was provoked by the position of the International Atomic Energy Agency (IAEA). For many years and particularly in the latest declaration (2005) [23], IAEA officials and experts have tried to persuade the world of their ideas or, more exactly, to demand (without convincing arguments) that the world community accepts their assurances about the safety of living within the regions contaminated by the Chernobyl accident as well as assurances relating to the "excessiveness" of measures taken by the Government on rendering adequate assistance to the population of the affected regions.

Every unprejudiced researcher, unbound by corporate interests, who has ever been to the places of residence of the Chernobyl-sufferers, especially in the Russian (Ukrainian, and Belarussian) remote regions, will consider the statement about the excessiveness of the measures taken as shocking. From the scientific point of view, it can be stated that there is no reliable evidence to be found for not only these mythical excessive measures but the provision of even the minimum necessary conditions for the safety of the people living in these regions – the victims of low-level irradiation. There is no adequate medical care or social support even intended for the improvement of these living conditions.

Perhaps, it is one of the most important lessons of the accident, which is the severest catastrophe in history, that any catastrophe is always irreversible, both for man and nature. Consequently, the question about the feasibility of safe living within the radiation-contaminated territories is senseless. It is relatively safe to live, but not in Chernobyl.

References

Akif'ev A.P., Obukhova L.K., and izmailov D.M., Vestnik Rus.Akad.Nauk., 1992, vol. 32, vol. 5, pp. 82-92. (in Russian)

Amundson S.A., Lee R.A., Koch-Paiz C.A., et al., *Differential Responses of Stress Genes to Low Dose Rate Irradiation,* Mol. Cancer Res., 2003, 1 (6), pp. 445-452.

Baleva L.S., Sipyagina A.P., in the book "20 Years after Chernobyl Catastrophe", 2006 (in press). I.V.

Burlakova E.B., Goloshchapov A.N., Gorbunova N.V., et al., Radiats. Biol. Radioecol., 1996, vol. 36, no. 4, pp. 610-631. (in Russian)

Burlakova E.B., Goloshchapov A.N., Gorbunova N.V., Gurevich S.M., Zhizhina G.P., et al., Radiats. Biol. Radioecol., 1996, vol. 36, no. 4, pp. 610-631. (in Russian)

Burlakova E.B., Goloshchapov A.N., Zhizhina G.P., and Konradov A.A.□*New aspects of effects of low doses of low-level irradiation*, Radiats. Biol. Radioecol., 1999, vol.32, no. 1, pp. 26-34. (in Russian)

Burlakova E.B., *Low Intensity Radiation: Radiobiological Aspects*, Rad. Protection Dosimetry, 1995, vol. 62, no. 1/2, pp. 13-18. (in Russian)

Burlakova E.B., *Some specific features of action of low dose irradiation*, Proceedings of the 24th Annual Meeting of the European Society of Radiation Biology, 1992, p. 88.

Burlakova E.B., Treshchenkova Yu.A., and Goloshchapov A.N., Radiats. Biol. Radioecol., 2003, vol. 43, no. 3, pp. 320-323. (in Russian)

Crompton N.E.A., Izsahin M., Schweizer P. et al., Strahlenther. Oncol., 1997, Bd. 2, S. 58.

Hardy K. and Stark J., *Mathematical models of the balance between apoptosis and proliferation*, Apoptosis, 2002, vol. 3, pp. 373-381.

Hooker A.M., Bhat M., Day T.K. et al., Rad. Res., 2004, no. 162, pp. 447-452.

Ivanov V.K., Chekin S.Y., Parshin V.S., et al., *Non-cancer thyroid diseases among children in the Kaluga and Bryansk regions of the Russian Federation exposed to radiation.*

M. Martin, F. Crechert, B. Ramount, and J-L. Lefaix, *Activation of c-fos by Low-Dose Radiation: Mechanism of the Adaptive Response in Skin Cells*, Radiat. Res., 141, 118 (1995).

Mil Å.Ì., Erokhin V.N., Kasparov V.V., et al., Biofizika, 2001, no. 46 (2), pp. 548-572. (in Russian)

Oradovskaya I.V., in the book "20 Years after Chernobyl Catastrophe", 2006 (in press).

OradovskayaI.V., in the book "20 Years after Chernobyl Catastrophe", 2006 (in press).

Pelevina I.I., Afanas'ev G.G.□, Gotlib V. Ya., and Serebryanyi A.B., in the book â êí. "Consequences of the Chernobyl accident: Human health" Ìoscow, Center of Ecol. Pol. 1996, pp. 229-244. (in Russian)

Pelevina I.I., Aleshchenko A.V., Gotlib V.Ya., et al., Radiats. Biol. Radioecol., 2004, vol. 44, no. 3, pp. 278-282. (in Russian)

Radiation and risk *Bulletin of the National Radiation-Epidemiology Register,* Moscow-Obninsk, 1996, no. 8. (in Russian)

Report of IAEA experts., Vienna, 2005.

Vartanyan L.S., Gurevich S.M., Kosachenko A.I., et al., *Age-related effects of low doses of ionizing radiation on the state of enzymic AO system of blood of participants of liquidation of consequences of the Chernobyl accident*, Uspekhi gerontologii, 2004, no. 14, pp. 48-54. (in Russian)

Yarilin A.A., Radiats. Biol. Radioecol., 1997, vol. 37, no. 4, pp. 597-603. (in Russian)

CHAPTER 3

Mental, Psychological and Central Nervous System Effects: Critical Comments on the Report of the UN Chernobyl Expert Group «HEALTH» (EGH)

Konstantin N. Loganovsky, MD, PhD, Dr. Med. Sci.

*Department of Radiation Psychoneurology, Institute for Clinical Radiology,
Research Centre for Radiation Medicine, Academy of Medical Sciences of Ukraine,
53 Melnikov Street, 04050, Kiev, Ukraine
Tel.: 380-44-452-1803; fax: 380-44-451-2330
e-mail: logan@rcrm.kiev.ua; psycho@rcrm.kiev.ua; psycho@ln.ua*

Background

International consensus exists that the mental health impact of theChernobyl accident is the greatest public health problem. The UN Chernobyl Forum Expert Group «Health» (EGH) has outlined four related areas of concern; stress-related symptoms; effects on the developing brain; organic brain disorders in highly exposed clean-up workers and suicide.

A heated scientific discussion over whether the Central Nervous System (CNS) is vulnerable to ionizing radiation has continued for more than a century. Both after the atomic bombing in Japan and, particularly, after the Chernobyl accident the world's research interests into radio-cerebral effects have significantly increased together with an augmentation of contradictions concerning brain radio-sensititvity. In spite of UNSCEAR-2000 (Annex J), whose experts acknowledged only the psycho-social consequences of the Chernobyl accident, the French IRSN agency identify the sensitivity of the CNS to low dose exposure: «…today it is recognized that the CNS is a radio-sensitive organ whose degree of dysfunction can be quantified by electrophysiological, biochemical and/or behavioural parameters. Abnormalities in CNS function defined by these parameters may occur at a low dose of whole body radiation…» (Gourmelon et al, 2005 — Institut de Radioprotection et de Surete Nucleaire, Fontenay-aux-Roses).

Undoubtedly, the clean up workers of the Chernobyl accident are under the highest risk of neuro-psychiatric disorders due to their greater exposure to both radiation and non-radiation factors of the disaster's aftermath. However, until now there is a gap in knowledge concerning the evidence-based estimation of their mental health. There are many contradictions concerning neuro-psychiatric effects at exposure to low doses (<1 Sv), e.g. whether such exposure is a risk factor for neuro-psychiatric disorders, particularly, schizophrenia spectrum disorders and Chronic Fatigue Syndrome (CFS) There are no clear data on cerebral radiation effects, cerebral radiation markers and dose–effect relationships.

In the currently available Working Draft, August 31, 2005 of the Report of the UN Chernobyl Forum EGH «Health Effects of the Chernobyl Accident and Special Health Care Programme» there are many gaps and even errors in the presentation of mental, psychological and CNS effects. The main goal of this paper is to discuss the currently available data and their limitations concerning the mental health aftermath of the Chernobyl accident, the neuro-psychiatric effects of ionizing radiation and investigate the research and measures for improving the mental health care of both the Chernobyl accident survivors and the casualties of possible radiation accidents in the future.

Stress-related disorders

Consensus exists concerning stress-related disorders following the Chernobyl accident. As it is presented in Chapter 15 of the Report of the UN Chernobyl Forum EGH, increased levels of depression, anxiety (including post-traumatic stress symptoms) and medically unexplained physical symptoms have been found in Chernobyl-exposed populations compared to controls (Viinamaki *et al*, 1995; Havenaar *et al*, 1997*a*; Bromet *et al*, 2000). Studies have also found that exposed populations had anxiety symptom levels that were twice as high and were 3-4 times more likely to report multiple unexplained physical symptoms and subjective poor health than the controls (Havenaar *et al,* 1997*b*; Allen and Rumyantseva; 1995; Bromet *et al,* 2002).

It was noted that these mental health consequences in the general population were mostly sub-clinical and did not reach the level of criteria for a psychiatric disorder (Havenaar *et al*, 1997*b*). Nevertheless these sub-clinical symptoms had important consequences for health behavior, specifically medical care utilization and adherence to safety advisories (Havenaar *et al*, 1997*a*; Allen and Rumyantseva, 1995).To some extent, these symptoms were driven by the belief that their health was adversely affected by the disaster and the fact that they were diagnosed by a physician as having a «Chernobyl-related health problem» (Bromet *et al*, 2002; Havenaar *et al,* 2003).

In the liquidator cohort (n=507) a very high level of mental disorders (84.42%) has been revealed. This is significantly higher than in inhabitants of radioactively contaminated territories of the Russian Federation (60.9%) and in populations of "clean" areas (47%). As opposed to populations where sub-clinical disorders are dominating, the proportion of clinically significant mental disorders among liquidators is quite large. Somatization and depression are especially pronounced (Rumyantseva *et al*, 1998).

Brain damage *in utero*

If the radio-vulnerability of an adult brain is accepted, it is axiomatic that the developing brain is extremely radiosensitive. Epidemiological studies on individuals who survived the atomic bombing of Hiroshima and Nagasaki and were exposed *in utero* confirm the vulnerability of the developing fetal brain to radiation injury. Severe mental retardation, lowering of intelligence quotient (IQ) and worsening of school performance, or the occurrence of microcephalia and seizures, especially after exposure at 8–15 and 16–25 weeks after fertilization were revealed (Otake and Shull, 1984, 1998; ICRP Publication 49, 1986; Shull, 1997; Shull and Otake, 1999). A recent re-analysis of the dosimetry data indicated that the dose threshold for the development of mental retardation after intra-uterine irradiation at gestation terms of 8–15 weeks is 0.06–0.31 Gy. At gestation terms of 16–25 weeks, it is 0.28–0.87 Gy (Otake *et al*, 1996). The question of increased lifetime prevalence of schizophrenia in survivors prenatally exposed to atomic bomb radiation is still open to discussion (Imamura *et al*, 1999).

The current concept about prenatal radio-cerebral effects is that 1 Sv of fetal exposure at 8–15 weeks of gestation reduces the IQ by 30 points. Correspondingly, it is assumed that each 100 mSv of prenatal irradiation lowers the IQ by no more than 3 points. The excess of severe mental retardation is 0.4 per 1 Sv at 8–15 weeks and, to a lesser extent, at 16–25 weeks of gestation (European Commission on Radiation Protection 100, 1998; ICRP Publication 84, 2000). Thus, the evidence-based radio-neuro-embriologic effects in humans are: 1) A dose-related intelligence reduction right up to mental retardation; 2) microcephalia; 3) seizures. This IQ deterioration depends upon the period of cerebrogenesis, when the exposure took place. However, other possible radio-neuro-embriological effects, such as schizophrenia and epilepsy are still being discussed.

Obviously, a great deal of concern has been expressed about the developing brain in children who were *in utero* when the accident occurred (Nyagu *et al*, 1993, 1996*a,b*, 1998, 2002*a*; Igumnov, 1996; Kolominsky *et al*, 1999; Kozlova *et al*, 1999; Loganovskaja and Loganovsky, 1999; Igumnov, Drozdovitch, 2000). On the one hand, the lowest level of exposure in which mental retardation was found in the offspring of survivors of Hiroshima and Nagasaki was higher than the highest level of exposure reported for most Chernobyl populations. On the other hand, there is a general belief that the brains of Chernobyl exposed children have been damaged. Thus, the World Health Organization conducted the Pilot Project on Brain Damage *In-Utero* within the framework of the International Program on the Health Effects of the Chernobyl Accident (IPHECA).

However, in spite of our numerous objections, on page 132 of Chapter 15 in the Report of the UN Chernobyl Forum EGH, Working Draft, August 31, 2005, a clear misinterpretation is still present as follows: «…Thus, the WHO conducted a pilot study of brain damage *in-utero* and *DID NOT find that exposed children had elevated rates of mental retardation compared to controls*». As a matter of fact, it must be as follows (World Health Organization, 1996, p. 402):

«…3. Analysis of the results of the investigations in the three countries has shown the following:
- Incidence of mental retardation in the lowest degree in the main group of children is higher when compared with the control group.
- An upward trend was detected in cases of behavioral disorders and in changes in the emotional problems in the children within the main group.
- Incidence of borderline nervous and psychological disorders in the parents of the main group is higher than that of the controls.

4. On the basis of these investigations it is impossible to arrive at a final conclusion on any relationship between *an increase in the number of mentally retarded children* and the ionizing radiation factor due to the Chernobyl accident. The results obtained are difficult to interpret and require verification. It is necessary to continue well planned epidemiological investigations and dosimetric follow-up of the project» (World Health Organization, 1996).

The principal conclusion that has been drawn on the basis of implementation of the IPHECA pilot project «Brain Damage *in Utero*» was the following: «*Arrested mental development and deviations in behavioral and emotional reactions is observed in a section of the children exposed in utero.* The contribution of radiation to such psychological changes still remains unclear due to the absence of individual dosimetry data»

(World Health Organization, 1996, p. 415).

Actually, as was presented in the EGH Report, two recent well-designed studies using standard batteries of neuropsychological tests failed to find systematic differences in children exposed *in utero* (Litcher *et al*, 2000; Bar *et al*, 2004). In the Litcher *et al.* study (2000), 31% of the mothers of evacuee children believed that their child had memory problems compared to 7% of controls even though there were no differences in actual neuro-psychological test performance or school grades. However, no dosimetric data were available, and there were no normative data in Ukraine for the measures used in the study (Bromet *et al*, 2000; Litcher *et al*, 2000). Moreover, the IQ tests were applied selectively: Litcher *et al.* (2000) assessed children's cognitive functions using spatial intelligence (Symbolic Relations subtest of the Detroit Test), attention and memory only and excluded verbal intelligence. A full scale IQ, verbal IQ and performance IQ are not available.

Using the ICD-10 criteria and additional assessments, Igumnov (1996) showed that mental disorders among children irradiated *in utero* resulted from predominantly socio-demographic and socio-cultural factors. An increased prevalence of specific developmental speech-language and emotional disorders, as well as a lower mean full scale IQ and more cases of borderline IQ in prenatally exposed children in Belarus were attributed to social and psychological factors. They did not find any association between prenatal irradiation and IQ and deterioration of mental health in children (Igumnov, 1999; Kolominsky *et al*, 1999). This point of view has been supported in the Annex J. Exposures and effects of the Chernobyl accident. UNSCEAR 2000 report to the General Assembly (2000). It was concluded that probably a significant role in the genesis of borderline intellectual functioning and emotional disorders in children exposed *in utero* in Belarus was due to unfavourable social-psychological and social-cultural factors (Igumnov & Drozdovitch, 2004).

For children exposed at the age of 0–1.5 years, there are data that show a correlation between thyroid dose (up to doses of 0.5 Gy) and reduction in intelligence (Bazyltchik *et al*, 2001). Average IQ for the subgroup of highly exposed children (thyroid doses *in utero* >1 Gy) was lower in comparison with average IQ for the whole exposed group (Igumnov and Drozdovitch, 2000). Prenatally irradiated children, especially those exposed at 8–15 weeks, had more functional and organic disorders of CNS, and exhibited borderline IQ and abnormal EEG linked to both radiation and psychosocial factors (Gayduk *et al*, 1994). Children irradiated *in utero* had the highest indices of mental morbidity and were more likely to display borderline intelligence and mental retardation that were linked to prenatal irradiation (Ermolina *et al*, 1996). The thyroid-stimulating hormone (TSH) level grows with foetal thyroid dose increase with the 0.3 Sv threshold. The radiation-induced malfunction of the thyroid-pituitary system was proposed as one important biological mechanism in the genesis of mental disorders in prenatally irradiated children (Nyagu *et al*, 1996*a,b*, 1998). It was hypothesized that malfunction of the left hemisphere limbic-reticular structures and the left hemisphere is more vulnerable to prenatal irradiation than the right (Loganovskaja and Loganovsky, 1999). The following were observed in prenatally exposed group; IQ performance/verbal discrepancies with verbal decrements; a higher frequency of low-voltage and epileptiformal EEG-patterns and left hemisphere lateralised dysfunction; an increase ($p<0.001$) of δ- and β-power and a decrease ($p<0.001$) of α- and θ-power; an increased frequency of paroxysmal and organic mental disorders, somatoform autonomic dysfunction, disorders of psychological development and behavioural and emotional disorders. Cerebral dysfunction was etiologically heterogeneous (Nyagu *et al*, 2002*a*).

Since possible dose correlations were not investigated and contradictory results of the mental health assessment of the *in utero* exposed children and the etiology of the observed neuropsychiatric disorders were found, a thorough study within the framework of Project 3 «Health Effects on the Chernobyl Accident» of the Franco-German Initiative for Chernobyl on potential effects of prenatal irradiation on the brain as a result of the Chernobyl accident has been carried out.

Neuro-psychiatric effects
It should be stressed that currently available data on radiation neuro-psychiatric effects in the Report of the UN Chernobyl Forum EGH, Working Draft, August 31, 2005 are still insufficiently presented, despite our constant, numerous on-line additions from the evidence-based medicine studies.

Irradiation in infancy and childhood

It has become increasingly clear that CNS radiotherapy in infancy and childhood may have serious long-term effects on cognition and endocrine function. As the treatment of childhood cancer has improved, long-term survival has become more common (Anderson, 2003). Currently, it is still assumed that the lowest dose on the brain that could be associated with late deterministic effects of childhood irradiation is 18 Gy, followed by disorders of cognitive functions, histo-pathological changes and neuro-endocrine effects (UNSCEAR 1993).

At the same time, evidence of delayed radiation brain damage was revealed 20 years after childhood scalp irradiation in average doses on the brain of only 1.3 Gy in a cohort of nearly 20,000 Israel children exposed to X-ray irradiation to the head for ringworm (tinea capitis) management (Yaar *et al.*, 1980, 1982; Ron *et al.*, 1982).

Recently, the effect of low doses of ionizing radiation (>100 mGy) in infancy (radiotherapy of cutaneous haemangioma) on cognitive function in adulthood has been proven on the basis of a Swedish population-based cohort study (Hall *et al*, 2004).

An increased risk of schizophrenia and related disorders was clearly seen among survivors who had been treated with radiotherapy for brain tumors in childhood or adolescence, as was shown in a nationwide, population-based, retrospective cohort study in Denmark (Ross *et al*, 2003).

Thus, radiation exposure in infancy and childhood is obviously associated with dose-related cognitive decline in adulthood. The possible dose thresholds of delayed radiation brain damage are doses as low as 0.1–1.3 Gy to the brain in infancy and childhood. However, the lowest fraction and total doses on the brain causing neurocognitive effects are still assumed to be 2 Gy and 18 Gy, correspondingly. Evidently further studies have to be done for reassessment of the risk-benefit of long-term consequences of cranial radiotherapy in infancy and childhood. Intellectual development is adversely affected when the infant brain is exposed to ionising radiation at doses (100 mGy) equivalent to those from computed tomography of the skull (120 mGy). The risk and benefits of computed tomography should be reassessed (Hall *et al.,* 2004).

Adult irradiation

Some deterministic radiation effects, which result in dysfunction of organs or tissue, are due not only to cell death. The dysfunction can be the result of interactions with functions of other tissues. Such «functional deterministic effects», in particular, include changes of electro-encephalogram (EEG) and retinogram, as well as vascular reactions. These effects can have significant clinical consequences, particularly in the nervous system (ICRP Publication 60, 1991). In spite of the fact that the mature CNS is commonly considered to be extremely radio-resistant, evidence is dramatically increasing in support of the exceptional radio-sensitivity of the brain (Nyagu and Loganovsky, 1998).

It is accepted in medical radiology that morphological radiation injuries of the CNS could arise following local brain irradiation by doses more than 10–50 Gy. Radiation brain necrosis was observed in local brain exposure to 70 Gy and more, where the development of radiogenic dementia is considered possible. The tolerant dose on the brain was assumed to be 55–65 Gy, and the tolerant fractional dose — 2 Gy (Gus'kova and Shakirova, 1989; Gutin *et al*, 1991; Mettler and Upton, 1995). Primary CNS damage following total body irradiation were assumed to be at exposure to >100 Gy (the cerebral form of Acute Radiation Sickness [ARS]), secondary radiation CNS damage — 50–100 Gy (the toxemic form of ARS) (Gus'kova and Bisogolov, 1971). The threshold of radiation-induced neuro-anatomic changes was assumed to be at the level of 2–4 Gy of whole body irradiation (Gus'kova and Shakirova, 1989).

However, in experimental studies, morphological changes of neurons were revealed as low as 0.25–1 Gy of total irradiation (Alexandrovskaja, 1959; Shabadash, 1964), and the dose of 0.5 Gy has been recognized to be the threshold of radiation injury of the CNS with primary neuronal damages (Lebedinsky and Nakhilnitzkaja, 1960). Persistent changes in brain bio-electrical activity occur at thresholds of 0.3 to 1 Gy and increase with the dose absorbed (Trocherie *et al.*, 1984). These data suggest that alteration in CNS functioning is likely to occur after relatively low doses of radiation (Mickley, 1987). It was shown that exposure to ionizing radiation significantly modifies neurotransmission (Kimeldorf and Hunt, 1965) and resulted in multiple effects on the brain and behaviour, depending largely on the dose received (Hunt, 1987). Ionizing radiation influences upon CNS function and behaviour as a result of both a direct effect to the nervous system and an indirect effect through CNS reactivity on the radiation damage of other systems (Kimeldorf and Hunt, 1965; Mickley, 1987). Slowly progressive CNS radiation sickness has been identified following a single exposure to total irradiation of 1–6 Gy (Moscalev, 1991). In the UNSCEAR Report (1982) it was noted that after exposure to 1–6 Gy a slow, progressive degeneration of the brain cortex develops (Vasculescu *et al*, 1973).

In the Adult Health Study in Hiroshima, atomic bomb radiation doses did not show any significant association with vascular dementia or Alzheimer's disease detected 25 to 30 years later. Risk factors for dementia were age, higher systolic blood pressure, history of stroke, history of hypertension, history of head trauma, lower milk intake and lower education (Yamada *et al*, 1999, 2003). However, taking into account that increased blood pressure was the main contributor to vascular dementia (Yamada *et al*, 1999), it is important that in the same Adult Health Study the statistically significant effect of ionizing radiation on the longitudinal trends of both systolic and diastolic blood pressure was recently found. This phenomenon is compatible with the degenerative effect of ionizing radiation on blood vessels (Sasaki *et al*, 2002). The recent analyses strengthen earlier findings of a statistically significant increase in non-cancer disease death rates with atomic bomb radiation dose. In particular, increasing trends are observed for diseases of the circulatory systems (Shimizu *et al*, 1999). There is direct evidence of radiation effects for doses more than 0.5 Sv for heart disease, stroke, digestive diseases and respiratory diseases (Preston *et al.*, 2003).

Epidemiological studies of atomic bomb survivors have suggested dose-related increases in mortality from diseases other than cancer. Cardiovascular disease is one such non-cancer disease for which increases in both mortality and incidence have been found to be associated with radiation dose (Kusunoki *et al.*, 1999). The recognition in the atomic-bomb survivors of non-cancer effects at doses on the order of 0.5 Sv (half the dose level considered a threshold in earlier studies) should stimulate interest in deterministic effects (Shimuzu *et al.*, 1999; Fry, 2001; Preston *et al.*, 2003; Yamada *et al.*, 2004) and non-cancer morbidity and mortality following the Chernobyl accident. It was indicated that the atomic bomb exposure has affected survivors' mental health and that the care of their mental health is important (Honda *et al.*, 2002). The prevalence of anxiety symptoms and somatization symptoms was elevated in atomic bomb survivors even 17-20 years after the bombings had occurred, indicating the long-term nature of the psychiatric effects of the experience (Yamada and Izumi, 2002). However, these studies were related to neurotic symptoms only — anxiety and somatization. At the same time, the linkage of The Life Span Study (LSS) and the schizophrenia register in the Department of Neuropsychiatry, University School of Medicine, Nagasaki revealed that the prevalence rate of schizophrenia in A-bomb survivors is very high — 6% (Nakane and Ohta, 1986).

Current estimates of lifetime prevalence of schizophrenia vary from 0.9–6.4 and an estimate of the mean prevalence is 1.4–4.6 per 1.000 (Jablensky, 2000). Included within the framework of the WHO 10-country study (Jablensky *et al*, 1992), the incidence of schizophrenia in Honolulu (USA) was 1.5 and in India 4.1 per 10.000 population (McGrath *et al*, 2004). The incidence of schizophrenia in India was the highest world-wide and this could not be explained solely by the level of psychiatric care (Tsirkin, 1987). In India there are areas of high natural radiation background due to monazite sands in Kerala, Madras, and the Ganges delta resulting in high average absorbed dose rate in the air of 1,800 nGy·h⁻¹ (UNSCEAR 2000). The coastal belt of the Trivandrum and Quilon districts of Kerala has a very high natural radioactivity — over 15 mSv per year (Rajendran *et al*., 1992). Worldwide annual exposures to natural radiation sources would generally be expected to be in the range 1–10 mSv, with 2.4 mSv being the present estimate of the central value (UNSCEAR 2000).

Since 1990, there has been a significant increase in schizophrenia incidence in the Chernobyl exclusion zone personnel (clean-up workers) compared to the general population (5.4 per 10,000 in the EZ versus 1.1 per 10,000 in the Ukraine in 1990) (Loganovsky and Loganovskaja, 2000).

Immediately after the Chernobyl accident autonomic [vegetative] vascular dystonia (VVD) and neurotic disorders were observed in the clinical picture of mild ARS [or ARS of the 1ˢᵗ severity degree] (0.8–2.1 Gy); moderate ARS [or ARS of the 2ⁿᵈ severity degree] (2–4 Gy) — VVD; severe ARS [or ARS of the 3ʳᵈ severity degree] (4.2–6.3 Gy) — acute radiation and radiation-toxic encephalopathy, acute psychosis with visual and acoustical hallucinations, brain edema; very severe to lethal ARS [or ARS of the 4ᵗʰ severity degree] (6–16 Gy) — acute radiation and radiation-toxic encephalopathy, subarachnoidal-parenchimatous hemorrhage, acute brain edema and swelling (Torubarov *et al*, 1989).

No clear signs of organic brain damage were registered during the first 3 years after irradiation in ARS-survivors. Mental working capacity deterioration and asthenisation were in proportion to the severity of ARS. The vegeto-vascular and vegeto-visceral stage of neuropsychiatric pathology (3–5 years after irradiation) changes with cerebral organic and somatogenous pathology (>5 years after irradiation) (Nyagu *et al*, 1996c, 2002b). Neuropsychiatric and neuropsychophysiological follow-up studies confirm that ARS-patients show progressive structural-functional brain damage: post-radiation encephalopathy [postradiation organic brain syndrome]. Even a considerable period of time after exposure quantitative EEG allows a distinction between confirmed and non-confirmed ARS (Nyagu *et al*, 2002b; Loganovsky, 2002).

The EEG-patterns and topographical distribution of spontaneous and evoked brain bioelectrical activity in overexposed liquidators, especially long-term workers of the Chernobyl zone, differed significantly from the control and comparison groups (Nyagu *et al*, 1992, 1999; Noshchenko and Loganovskii, 1994; Loganovsky, 2000). There were many consistent reports concerning characteristic neurophysiological disorders (Danilov and Pozdeev, 1994; Zhavoronkova *et al*, 1994, 1995a,b, 1998; Vyatleva *et al*, 1997), neuropsychological (Khomskaja, 1995; Zhavoronkova *et al*, 1996, 2000), and neuroimaging (Zhavoronkova *et al*, 1994; Kharchenko *et al*, 1995; Kholodova *et al*, 1996; Voloshina, 1997) abnormalities in liquidators, supporting clinical data about organic brain damage (Chuprikov *et al*, 1992; Krasnov *et al*, 1993; Romodanov and Vynnyts'kyj, 1993; Napreyenko and Loganovsky, 1995, 1997, 1999, 2001a,b; Nyagu and Loganovsky, 1998; Revenok, 1998, 1999; Zozulya *et al*, 1998; Morozov and Kryzhanovskaja, 1998; Lysyanyj, 1998).

The progressive character of neuropsychiatric disorders and somatic pathology is observed in liquidatiors exposed between 1986-1987, especially in those who worked for 3-5 years within the Chernobyl exclusion zone. A prevalence of neuropsychiatric disorders was observed among personnel working between 1986-1987 and irradiated in doses above 250 mSv of 80.5% while, for the same contingency, but irradiated in doses below 250 mSv – 21.4% (p<0.001) (Bebeshko *et al*, 2001; Nyagu *et al*, 2003). Personnel who have been working within the Chernobyl exclusion zone since 1986 are the group at highest risk of neuropsychiatric disorders, where organic, including symptomatic, mental disorders (F00— F09) dominate (Loganovsky, 1999). Those irradiated by moderate to high doses (more than 0.3 Sv), including ARS patients, had significantly more left fronto-temporal limbic dysfunction and schizophreniform syndromes (Loganovsky, 2000a; Loganovsky and Loganovskaja 2000). Among a considerable part of the personnel, especially those who carried on with their work into the 1990-s, the pathology revealed met the criteria of Chronic Fatigue Syndrome (CFS). A hypothesis was suggested concerning development of CFS under impact of low and very low doses combined with psychological stress (Loganovsky *et al*, 1999, Loganovsky, 2000*b*).

These findings should be confirmed by international collaborative studies. Moreover, the biological basis of the relationships should be revealed.

Suicides

As was presented at the Report of the UN Chernobyl Forum EGH, suicide was the leading cause of death among Estonian clean-up workers (Rahu *et al*, 1997). Age-adjusted mortality from suicide rates were higher among the Chernobyl clean-up workers compared to the general population in Lithuania (Kesminiene *et al*, 1997). These findings have to be replicated in studies of clean-up workers from other countries using standardized methodology for suicides due to possible misclassification of them as another cause of death.

Summary

It is true, as outlined in the Report of the UN Chernobyl Forum EGH, that the context in which the Chernobyl accident occurred, i.e., the complicated series of stressful events unleashed by the accident, the self-fulfilling consequences of the official label «Chernobyl victim» given to the affected population, the multiple stress factors that occurred in the former Soviet Union before and after Chernobyl and the culture-specific ways of expressing distress, makes the findings difficult to interpret precisely.

At the same time, an increasing pool of data on radiation cerebral effects cannot be ignored. This concerns;

- Non-cancer effects in atomic-bomb survivors at doses of 0.5 Sv (Shimuzu *et al*, 1999; Preston *et al*, 2003);
- Epidemiological evidences on dose-related cognitive decline following radiotherapy in childhood with possible dose thresholds of delayed radiation brain damage at doses as low as 0.1–1.3 Gy on the brain (Yaar *et al*, 1980, 1982; Ron *et al*, 1982; Hall *et al*, 2004);
- Cognitive and emotional-behavioral disorders (WHO, 1996; Nyagu *et al*, 1998, 2002*a*; Kolominsky *et al*, 1999; Loganovskaja and Loganovsky, 1999; Igumnov and Drozdovitch, 2000), and neurophysiological abnormalities (Nyagu *et al*, 1996*a,b*, 1998, 2002*a*; Loganovskaja and Loganovsky, 1999) in prenatally exposed children;
- Postradiation organic brain syndrome in ARS-patients (Nyagu *et al*, 2002*b*; Loganovsky, 2002) and characteristic neurophysiological, neuropsychological and neuroimaging abnormalities in liquidators, supports clinical neuropsychiatric data about organic brain damage where, following exposure from moderate to high doses (more than 0.3 Sv) left fronto-temporal limbic

dysfunction and schizophreniform syndromes dominate (Loganovsky, 2000*a*, 2002; Loganovsky and Loganovskaja, 2000);

- Schizophrenia and related disorders (Nakane and Ohta, 1986; Imamura *et al*, 1999; Loganovsky and Loganovskaja, 2000);
- Chronic Fatigue Syndrome following exposure to low and very low doses combined with psychological stress (Loganovsky *et al*, 1999, Loganovsky, 2000*b*).
- Obviously, these effects associated with exposure to ionizing radiation have to be confirmed by international collaborative studies. Moreover, their biological basis should be elucidated.

Current status of evidence

Potential prenatal cerebral effects

Extrapolation of the Japanese data to the situation after the Chernobyl accident is quite difficult. Thus, the Chernobyl accident caused significantly lower foetal doses, but significantly much higher doses on the foetal thyroid by the incorporation of radioiodine released by the burning reactor. Whereas after the Chernobyl accident the population was continuously exposed to radionuclides, mainly of radioiodine and ^{137}Cs, the Japanese population was acutely irradiated by γ-rays and neutrons. There was no separate radioiodine exposure of the thyroid in Japan. Because of the different radiobiological situations, it is not easy to predict the radiobiological effect of the Chernobyl accident from the results of the Japanese studies (Nyagu *et al*, 2004*b*).

Due to the contradictory results of the mental health assessments of the children exposed *in utero* and the etiology of the observed neuropsychiatric disorders in the literature, a thorough study of the potential radiation effects on the mental health of the *in utero* exposed children was performed within the framework of Project 3 «Health Effects on the Chernobyl Accident» of the French-German Initiative for Chernobyl. A cohort of 154 children born between April 26th, 1986 and February 26th, 1987 to mothers who had been evacuated from Pripyat to Kiev, and 143 classmates from Kiev were examined with the Wechsler Intelligence Scale for Children (WISC) and the Achenbach and Rutter A(2) tests. Mothers were tested for their verbal abilities (WAIS), depression, anxiety and somatization (SDS, PTSD, GHQ 28). Individual dose reconstruction of the children was carried out taking into account internal and external exposure. The ICRP Publication-88 was applied for calculation of effective foetal, brain and thyroid internal doses for children of both groups. There were 52 children from Pripyat (33.8%) who had been exposed *in utero* to equivalent thyroid doses >1 Sv, 20 of these children 13.2% had been exposed *in utero* to fetal doses >100 mSv. The prenatally exposed children showed significantly more mental disorders and diseases of the nervous system. Exposed children showed lower full-scale IQ due to lower verbal IQ and therefore an increased frequency of performance/verbal intelligence discrepancies. When IQ discrepancies of the prenatally irradiated children exceeded 25 points, there appeared to be a correlation with the foetal dose. The exposed and control mothers did not show differences in verbal abilities, but they had experienced many more real stress events and suffered from severer depression, PTSD, somatoform disorders, anxiety/insomnia and social dysfunctions than the control mothers from Kiev (Nyagu *et al*, 2004*a,b,c*).

A radio-neuro-embriological effect — intelligence disharmony due to verbal IQ deterioration — has been revealed following the radiation accident at the nuclear reactor at 8–15 and later weeks of gestation, fetal >20 mSv and thyroid doses *in utero* >300 mSv. Spectral θ-power decreased (particularly, in the left fronto-temporal area), β-activity increased together with its lateralization towards the dominant hemisphere, disorders of normal inter-hemispheric asymmetry of visual evoked potentials and the vertex-potential can be considered as neurophysiological markers of prenatal exposure. The most critical period of cerebrogenesis occurs during the later terms of gestation (16–25 weeks). The radiation accident on the nuclear reactor resulted in a release of radioactive iodine into the environment and this had a greater effect on cerebrogenesis than the uniform external exposure (Loganovskaja and Nechayev 2004; Loganovskaja, 2004, 2005). The frequency of mental disorders and personality disorders due to brain injury or dysfunction — F06, F07; disorders of psychological development — F80–F89; paroxysmal states (headache syndromes — G44, migraine — G43, epileptiform syndromes — G40); somatoform autonomous dysfunction — F45.3; behavioral and emotional disorders of childhood — F90–F99 were increased among these children (Napreyenko and Loganovskaja, 2004; Loganovskaja, 2005).

A follow-up study is currently underway, funded by the United States National Institute of Health (NIH), to examine the health and mental health of a cohort of 18-19 year olds who were *in utero*–15 months old when the accident occurred and aged 11 at the time of first assessment. The study addresses the extent to which initial perceptions of the event are risk factors for future mental health, as well as the differences at age 18 between exposed children and their mothers versus the original classmate control group in *Kiev* and a new population-based control group from other unaffected urban areas in Ukraine.

Population studies

It was concluded that epidemiological data do not at present provide clear evidence of a risk of circulatory diseases at doses of ionizing radiation in the range 0-4 Sv, as suggested by the atomic bomb survivors. Further evidence is needed to characterize the possible risk (McGale and Darby, 2005).

Within the framework of the Franco-German Chernobyl Initiative sub-project 3.8.1 «Data base on psychological disorders in the Ukrainian liquidators of the Chernobyl accident» A cross-sectional study of a representative cohort of liquidators, using a standardized structured psychiatric interview — Composite International Diagnostic Interview (CIDI), was carried out. The preliminary results have been revealed (Romanenko *et al*, 2004): a two-fold increase in the prevalence of mental disorders (36%) in liquidators in comparison with the Ukrainian general population (20.5%); a dramatic increase in the prevalence of depression (24.5%) in liquidators in comparison with the Ukrainian general population (9.1%) (Demyttenaere *et al*, 2004). The dataset is open to analysis. However, the limitation of this study is the assessment of psychological disorders only. It excludes severe mental disorders.

Compared with controls, the evacuees reported significantly more health problems and rated their health more poorly overall. The relationship between Chernobyl stress and illness was twice as strong in evacuees (odds ratio = 6.95) as in *Kiev* controls (odds ratio = 3.34) and weakest in the national sample (odds ratio = 1.64). The results confirm the persistence and nonspecificity of the subjective medical consequences of Chernobyl and are consistent with the hypothesis that traumatic events exert their greatest negative impacts on health in vulnerable or disadvantaged groups (Adams *et al*, 2002; Bromet *et al*, 2002).

The progressive character of neuropsychiatric disorders and somatic pathology is observed in the liquidatiors of 1986-1987, especially in those who worked for 3-5 years within the Chernobyl exclusion zone. Prevalence of neuropsychiatric disorders among personnel working since 1986-1987 and irradiated in doses above 250 mSv is 80.5%, while for the same contingency but irradiated in doses below 250 mSv – 21.4% ($p<0.001$) (Nyagu *et al*, 2003).

It has been revealed that since 1990 there has been an increase in the incidence rate of schizophrenia in the personnel compared with the general population: 5,4 per 10.000 vs. 1,1 per 10,000 correspondingly. Relative risks are 2,4 in 1986—97 and 3,4 in 1990—97, testifying to an association between working and living in the Chernobyl exclusion zone and an increase in the schizophrenia incidence rate by 2,4—3,4 times in comparison with the general population.

A significant increase has also been established of schizophrenia percentage among all psychoses among the Chernobyl exclusion zone staff members in comparison with the general population. Moreover, a non-typical clinical form of schizophrenia was revealed in the personnel who had been working since 1986—87. Those irradiated by moderate to high doses (more than 0.3 Sv) had significantly more left fronto-temporal limbic and schizophreniform syndromes. It has been hypothesized that ionizing radiation may be an environmental trigger that can actualize a predisposition to schizophrenia or indeed cause schizophrenia-like disorders (Loganovsky and Loganovskaja, 2000). An integration of international efforts to discuss and organize collaborative studies in this field is of significance for both clinical medicine and neuroscience (Loganovsky et al, 2005) In the Chernobyl accident clean-up workers (liquidators) the non-cancer effects of radiation risks has been revealed (Biriukov et al, 2001; Buzunov et al, 2001, 2003). For some classes of non-cancer diseases among liquidators, statistically significant estimates of radiation risk were derived for the first time: mental disorders — ERR 1/Gy=0.4 (0.17; 0.64); neurological and sensory disorders — ERR 1/Gy=0.35 (0.19; 0.52); endocrine disorders — ERR 1/Gy=0.58 (0.3; 0.87). Among mental disorders the higher radiation risks were revealed for neurotic disorders— ERR =0.82 (0.32; 1.32) (Biriukov et al, 2001). The highest excess relative risk per 1 Gy was found for cerebrovascular diseases — ERR 1/Gy=1.17 at the 95% confidence interval (0.45; 1.88) (Ivanov et al, 2000). Recently, the significant risk of cerebrovascular diseases from an average dose rate was defined for external doses greater then 150 mGy (ERR for 100 mGy/day = 2.17, with 95% CI = (0.64; 3.69) (Ivanov et al, 2005).

According to the data from the State Register of Ukraine and the Clinical and Epidemiological Registry (Scientific Centre for Radiation Medicine, Kiev) there is an increased level of cerebrovascular disorders in liquidators and evacuees. Exposure to small doses of ionizing radiation is a significant risk factor of accelerating ageing. Thyroid exposure by 300 mGy and more is a significant risk factor for vascular and cerebrovascular disorders. Thyroid exposure by 2 Gy or more is a significant risk factor for mental disorders, vascular and cerebrovascular diseases, and the peripheral nervous system. Exposure to dose 250 mGy or more is a significant risk factor for neuropsychiatric disorders and vascular disorders. There is a dose–effect relationship for cerebrovascular disorders in liquidators. Relative risk of cerebrovascular diseases is increased in the groups exposed to 0.5–0.99 Gy and 1 Gy in comparison with the group of <0.1 Gy. Non-radiation risk factors for neuropsychiatric pathology (cerebrovascular) includes: industrial hazards, stress, smoking, heredity and life style (Buzunov et al, 2001, 2003).

However, concerning the mental health assessment of liquidators, these studies have significant limitations: they deal with mental disorders registered by the national health care system, but not with the data obtained as a result of well-designed psychiatric studies with standardized diagnostic procedure. Taking into account also the current changes of the psychiatric system in the post-soviet countries, this leads to a dramatic underestimation of mental disorders and their possible misclassification as physical diseases and/or wrong diagnoses of mental disorder itself (neurotic as opposed to actually psychotic or organic, etc.).

For example, according to the official Public Health Ministry of Ukraine data, the prevalence of all mental disorders (registered) in the Ukrainian population consisted: in 1990 — 2.27%; in 1995 — 2.27%, and in 2000 — 2.43%. However, according to the results of the World Health Organization (WHO) World Mental Health (WMH) Survey Initiative, where they assessed mental disorders with the WMH version of the WHO Composite International Diagnostic Interview (WMH-CIDI), a fully structured, lay-administered psychiatric diagnostic interview, the prevalence of having any WMH-CIDI/DSM-IV disorder in the prior year in Ukraine was 20.5% (95% CI —17.7–23.3%) (Demyttenaere et al, 2004). So, the public health psychiatric system underestimates mental disorders by at least 10 times. It should be stressed, that the WMH-CIDI/DSM-IV disorders included so-called psychological disorders (anxiety, depression, somatization, alcohol abuse, etc.) only and did not deal with severe mental disorders — psychoses, organic mental disorders, and mental retardation.

Neuropsychiatric effects

A prospective conventional EEG study was carried out 3-5 and 10-13 years after the Chernobyl accident in patients who had ARS and in liquidators of 1986. Control groups comprised healthy volunteers; veterans of the Afghanistan war with post-traumatic stress disorder; veterans with mild traumatic brain injury; and patients with dyscirculatory encephalopathy. Within 3-5 years after irradiation, there were irritated EEG changes with paroxysmal activity shifting to the left fronto-temporal region (cortical-limbic over activation) that were transformed 10-13 years after irradiation toward a low-voltage EEG pattern with excess of fast (beta) and slow (delta) activity together with depression of alpha and theta activity (organic brain damage with inhibition of the cortical-limbic system) (Loganovsky and Yuryev, 2001).

Cross-sectional quantitative electroencephalogram (qEEG) studies (1996-2001) among Chernobyl accident survivors, who had confirmed ARS and were irradiated in doses of 1-5 Gy, revealed the neurophysiological markers of ionizing radiation. Neurophysiological markers were: left fronto-temporal dominant frequency reduction; absolute delta-power lateralization to the left (dominant) hemisphere; relative delta-power increase in the fronto-temporal areas; absolute theta-power decrease in the left temporal region; absolute and relative alpha-power diffusive decrease, which may reflect cortico-limbic dysfunction lateralized to the left, dominant hemisphere, with the fronto-temporal cortical and hippocampal damage. Quantitative electroencephalogram proposed for differentiation of radiation and nonradiation brain damages and as a new biological dosymetry method. High radiosensitivity of the brain, neocortex, and dominant hemisphere higher radiosensitivity are discussed. (Loganovsky and Yuryev, 2004).

Changes in brain asymmetry and inter-hemispheric interaction can be not only a result of a dysfunction of subcortical limbic-reticular and mediobasal brain structures but also a result of the white matter damage including corpus callosum (Zhavoronkova et al, 2000). The EEG findings suggest subcortical disorders at different levels (diencephalic or brainstem) and functional failure of the right or left hemispheres in remote terms after exposure to radiation (Zhavoronkova et al, 2003).

A 4-year longitudinal study of the cognitive effects of the Chernobyl nuclear accident was conducted from 1995 to 1998. The controls were healthy Ukrainians residing several hundred kilometers away from Chernobyl. The exposed groups included liquidators, Forestry workers and Agricultural workers living within 150 km of Chernobyl. Accuracy and efficiency of cognitive performance were assessed using ANAMUKR, a specialized subset of the Automated Neuropsychological Assessment Metrics (ANAM) battery of tests. Analyses of variance, followed by appropriate pair-wise comparisons, indicated that the 4-year average levels of performance of the exposure groups (especially the liquidators) were significantly lower than those of the controls in most measures; further analyses of performance across time revealed significant declines in accuracy and efficiency, as well as psychomotor slowing, for all exposed groups over the 4-year period. These findings strongly indicate impairment of brain function resulting from both acute and chronic exposure to ionizing radiation (Gamache et al, 2005).

The neural diathesis-stressor hypothesis of schizophrenia, where neurobiological genetic predisposition to schizophrenia can be provoked by environmental stressors is considered as a model of the effects of exposure to ionizing radiation. There are comparable reports on increases in schizophrenia spectrum disorders following exposure to ionizing radiation as a result of atomic bombing, nuclear weapons testing, the Chernobyl accident, environmental contamination by radioactive waste, radiotherapy, as well as in areas with high natural radioactive background (Loganovsky et al., 2004a, 2005a). The results of experimental radioneurobiological studies support the hypothesis of schizophrenia as a neurodegenerative disease (Korr et al, 2001; Gelowitz et al, 2002; Schindler et al, 2002). Exposure to ionizing radiation causes brain damage with cortical-limbic system dysfunction and impairment of informative processes at the molecular level that can trigger schizophrenia in predisposed individuals or cause schizophrenia-like disorders (Loganovsky et al., 2004a, 2005a).

Negative psychopathological symptoms increase in proportion to the radiation dose (at doses >0.3 Sv), but symptoms of PTSD has an inverse relationship to the dose. Depression is more pronounced among the patients who had the severest ARS. After irradiation at doses >0.3 Sv and, especially after ARS (>1 Sv), there are a predominant involvement of the fronto-temporal regions of the dominating (left) hemisphere; EEG slowing; progressive decreasing of absolute spectral power («flat» low-voltage EEG) with increased δ- and β-power and decreased θ- and α-power. Apathetic-abulic endoformous organic brain syndrome prevails in liquidators exposed to 0.3–0.5 Sv and more while cerebrasthenic syndrome or cerebrasthenic and dysthymic variants of organic brain syndrome predominate at less 0.3 Sv. The relationship between neurophysiological effects and radiation dose was revealed only in doses more than 0.3 Sv - that proposed as the threshold of these effects. This dependence is significantly intensified in long-term work (more than 5 years) within the Chernobyl exclusion zone (Loganovsky, 2000a, 2002).

Post-radiation organic brain syndrome has been diagnosed in 62% of patients who had confirmed ARS. The apathetic type of organic personality disorder (F07.0) is a characteristic for ARS after effects. It has a progressive clinical course and correlates with the severity of ARS or the dose (Loganovsky, 2002; Loganovsky et al., 2005b). Organic brain damage, long after ARS, has been verified by clinical neuropsychiatric, neurophysiolocial, neuropsychological and neuroimaging methods (Loganovsky et al, 2003; 2005b).

There are neuropsychological dose-effects relationships in humans at 0.5-5 Sv dose range — dysfunction of memory interferention (r=0.6); dysfunction of associations (r=0.68), and dysfunction of senso-motor coordination (r=0.8) (Antipchuk, 2004, 2005). The cerebral effects, long after irradiation, are a pathology of frontal and temporal cortex of the dominant hemisphere and middle structures with their cortical-sub-cortical connections, resulting in higher mental activity deterioration in structural, organic and psychical disorders (Antipchuk, 2001, 2003, 2005). At the same time, memory, acoustic-perceptual and, partially, thinking functions were most vulnerable for liquidators exposed below 0.35 Sv (Turuspekova, 2002). Dose—effect relationship on mental working capacity deterioration was revealed for the ARS patients (>1 Sv) (Zdorenko and Loganovsky, 2002). The reduction of verbal and full IQ in comparison with controls, as well as IQ deterioration according to pre-exposure IQ estimates, are evidence of organic brain syndrome (with paramount involvement of the left, dominating, hemisphere) following ARS (Loganovsky and Zdorenko, 2004).

In spite of the fact that the CNS is commonly considered to be extremely radioresistant, evidence is dramatically increasing in support of the exceptional radiosensitivity of the brain. There is the «dose–effect» relationship between the dose and the characteristic morphometric neuroimaging features of organic brain damage, starting with 0.3 Sv and increasing in proportion to the dose. There is the neuroimaging dose-effects relationship in humans by brain MRI considered to be neuroimaging radiation markers. They are at 1–5 Sv: contrast index of left internal capsule (r=-0.86) and contrast index of white matter of left parietal lobe (r=-0.41); the markers at 0.3–5 Sv: contrast index of left internal capsule (r=-0.3). These data testify to radiation-induced organic brain damage among over-exposed Chernobyl accident victims and support the presence of radiation risks for the brain as well as high radiosensitivity of the brain (Bomko, 2004a, 2005). Cortical atrophy of cerebral hemispheres and damage of neuronal pathways in the dominant hemisphere are the characteristic morphometric neuroimaging features of organic brain damage to ionizing radiation as a result of the Chernobyl accident (Bomko, 2004b, 2005).

Brain damage following earlier exposure to irradiation as a result of the Chernobyl accident has characteristic patterns testifying to organic damage of gray and white brain matter of atrophic genesis with predominant involvement of the frontal and temporal lobes, particularly, of the left (dominating) hemisphere (Loganovsky et al, 2004b). The structural-functional pattern of radiation-induced organic brain damage in Chernobyl accident clean-up workers consists in the pathology of cerebral cortex, subcortical structures, neuronal pathways and cortical-limbic system of the left (predominantly) cerebral hemisphere. Associations between neurophysiological and neuroimaging parameters and dose make dose reconstruction possible using parameters of cEEG and brain MRI long after exposure to ionizing radiation (Loganovsky and Bomko, 2004).

Radiation accelerated aging might be a model of senescence and neurodegeneration in humans. Prospective epidemiological studies of atomic bomb survivors revealed ionizing radiation significantly increased mortality for causes other than cancer. Results do not support claims that survivors exposed to low doses live longer than comparable unexposed individuals. Whether exposure to a low dose is a risk factor for accelerated aging and neurodegeneration is still unanswered and the biological mechanisms involved unknown. According to a wide range of scientific data reviewed, the following hypotheses can be proposed: 1) exposure to low-dose ionizing radiation is a risk factor for accelerated aging processes and neurodegeneration; 2) aging and neurodegeneration processes after exposure to ionizing radiation could be enhanced by the synergetic influence of heterogeneous pathogenetic factors, such as immunological, oxidative stress and molecular-genetic changes (Bazyka et al, 2004).

From 100 randomly selected clean up workers who were involved with the Chernobyl accident 26% met the CFS diagnostic criteria. Their absorbed doses were less 0.3 Sv, which could not produce any clear deterministic radiation effect. A psycho-physiological basis for fatigue in liquidators is dysfunction of the cortico-limbical structures of the left, dominating, hemisphere. CFS is one of the most significant consequences of radio-ecological disaster resulting in an interaction of different hazardous environmental factors (Loganovsky, 2000b, 2003). CFS frequency significantly (p<0,001) decreased (from 65,5% in 1990–1995 to 10,5% in 1996–2001) and Metabolic Syndrome X (MSX) frequency significantly (p<0,001) increased (from 15 to 48,2%). CFS and MSX are considered to be early stages of other neuropsychiatric and physical pathology development, and CFS can transform towards MSX. Radiation-induced damage of mitochondrial DNA in post-mitotic tissues with low proliferation activity may be one of the effects of low doses contributing to an increase in non-cancer morbidity and mortality in the Chernobyl accident survivors (Kovalenko and Loganovsky 2001). CFS can be considered an environmentally induced predisposition and precursor of forthcoming neurodegeneration, cognitive impairment, and neuropsychiatric disorders. The generalized conceptual framework unique to the Chernobyl accident data, including clinical, molecular biomarkers, and different exposure to multiple environmental stress-factors (acute and chronic exogenous and endogenous (radionuclides) irradiation, chemical agents, viruses, severe psychological stress, etc) allow us to investigate, conceptualize, and illustrate fundamental aspects of CFS etiology and pathophysiology and its neurodegeneration and neuropsychiatric perspectives (Volovik *et al*, 2005).

Interventions

The attentions of The Radiation Centre for Radiation Medicine (*Kiev*) were constantly focused on the improvement of the mental health of the affected population. The lessons to be learned from the Chernobyl accident experience include;

1) Neuropsychiatric care in the aftermath of ARS and overexposure to ionising radiation should be a priority area of attention following exposure to accidental irradiation;

2) Neuropsychiatric treatment should be carried out as early as possible, as with brain trauma;

3) Limitations should be imposed on duration of work in the Chernobyl exclusion zone (and similar industries) in order to provide adequate mental health care of personnel. At the end of this period working staff members should terminate their work or begin rehabilitation. A 5-years work-limit is recommended;

4) Effective protection of mental health is only possible by simultaneously addressing the neuropsychiatric, personal, somatic and social spheres of the lives of the survivors. In this regard, the establishment of a National Service for Mental Health for the protection of survivors should be a priority;

5) A social measures system is needed;

6) Non-pharmacological interventions;

7) Pharmacological treatment of main and co-morbid pathology;

8) Long-term support and maintenance treatment.

Conclusions

1) Mental disorders are one of the most important medical and social problems facing Chernobyl accident survivors 20 years after the Chernobyl accident

2) Mental health care of victims should be a focus of public concern following possible radiation accidents in the future;

3) Mental health impacts on the liquidators of the Chernobyl accident includes:
   - psychological disorders
   - organic brain damage
   - suicides
   - chronic Fatigue Syndrome
   - schizophrenia spectrum disorders
   - accelerated aging processes and neurodegeneration

4) There is a big gap in epidemiological evidence concerning the mental health of exposed populations, as well as a gap in knowledge about the biological mechanisms of the effects of low doses on the brain

5) There is a deficiency in mental health care and psycho-rehabilitation for survivors.
6) According to current knowledge, the potential cerebral effects of radiation could be outlined as follows:
   - Potential cerebral effects of radiation could be realized following exposure to >0.15–0.5 Sv
   - Dose-related cognitive decline following radiotherapy in childhood with possible dose thresholds of delayed radiation brain damage at doses as low as 0.1–1.3 Gy on the brain;
   - Dose-related cognitive and neurophysiological abnormalities in prenatally exposed children;
   - Postradiation organic brain syndrome in ARS-patients and dose-related neuropsychiatric, neurophysiological, neuropsychological and neuroimaging abnormalities following exposure to >0.3 Sv;
7) Effects on the CNS that could be attributed to exposure of ionizing radiation are as follows:
   - Schizophrenia spectrum disorders;
   - Chronic Fatigue Syndrome;
   - Accelerated aging processes and neurodegeneration.

Recommendations

1) Further well-designed neuropsychiatric epidemiological studies in Chernobyl accident survivors should have a priority. An integration of international efforts to discuss and organize collaborative studies concerning neuropsychiatric disorders, including organic brain damage, CFS, schizophrenia spectrum disorders, suicides and para-suicides are of great importance for both clinical medicine and neuroscience.

2) The development of a model which will predict safety levels in the case of extreme environmental stress factors associated with radiation exposure using the data from the Chernobyl accident.

3) Neuropsychiatric follow up survey of ARS-patients must be continued for a life time. *In vivo* morpho-functional (fMRI; qEEG, EP) and neurochemical studies (SPET) are of the greatest priority.

4) The study of cerebral effects of prenatal exposure should be continued as follows; an increase in the size of the cohort; identification of additional children irradiated *in utero* and children exposed at the age of 0–1 years; the identification and forming of cohorts of age-, gender- and urban/rural-matched children from radioactively clean areas of the Ukraine; the verification and development of the currently available dosimetric models; the assessment and verification of neuropsychiatric disorders and a full risk analysis should be carried out on the influence of radioiodine on brain development during the prenatal period and the 1st year of life.

5) Further clinical and experimental neurophysiological, neurobehavior, neuroimaging, neurochemical and neuroimmunological studies are of the greatest importance for brain assessment and radiation risks.

6) Neuropsychiatric care in the aftermath of ARS and overexposure to ionising radiation should be a priority area of attention and this treatment should be carried out as early as possible, as with any trauma to the brain;

7) Maximum work duration in the Chernobyl exclusion zone (and similar industries) should be limited to 5-years;

8) International efforts should be made to improve the mental health care and psycho-rehabilitation of Chernobyl accident survivors.

References

Adams RE, Bromet EJ, Panina N, Golovakha E, Goldgaber D, Gluzman S (2002) Stress and well-being in mothers of young children 11 years after the Chornobyl nuclear power plant accident. *Psychol Med.* 32(1):143–156.

Alexandrovskaja MM (1959) Effects of different doses of ionizing radiation on the brain morphology in animals at total irradiation. *Medical Radiology*, 4 (8): 79–81 [in Russian].

Allen PT, Rumyantseva G (1995). The contribution of social and psychological factors to relative radiation ingestion dose in two Russian towns affected by the Chernobyl NPP accident. *Society for Risk Analysis (Europe)*, p 1-9

Anderson NE (2003) Late complications in childhood central nervous system tumour survivors. *Curr Opin Neurol*, 16(6): 677–683.

Antipchuk YeYu (2001) Disorders of the highest cortical functions in clean-up workers (1986) of the Chernobyl accident with encephalopathy. In: *Problems of Radiation Medicine*, Research Centre for Radiation Medicine of Academy of Medical Sciences of Ukraine. *Kiev*, Vol. 8, pp 83–89 [in Russian].

Antipchuk KYu (2003) Memory disorders in persons who had acute radiation sickness as a result of the Chernobyl accident in remote period. *Ukrainian Radiological Journal,* 11: 68–72 [in Ukrainian].

Antipchuk KYu (2004) Neuropsychologic method in diagnostic of radiation brain disorders. *Ukrainian Medical Journal*, 3(41): 121–128 [in Ukrainian].

Antipchuk KYu (2005) *Clinical-neuropsychological characteristic of organic mental disorders in remote period of exposure to ionizing radiation as a result of the Chernobyl accident*. The dissertation for the academic degree of a Candidate of Medical Sciences in radiobiology. Research Centre for Radiation Medicine of Academy of Medical Sciences of Ukraine. *Kiev*.

Bar Joseph N, Reisfeld D, Tirosh E, Silman Z, Rennert G. (2004) Neurobehavioral and cognitive performance in children exposed to low-dose radiation in the Chernobyl accident: The Israeli Chernobyl Health Effects Study. *Am J Epidemiol, 160*, 453-459

Bazyka DA, Volovik SV, Manton KG, Loganovsky KN, Kovalenko AN (2004) Ionizing radiation accelerating aging and neurodegeneration. *International Journal of Psychophysiology*, 54 (1–2): 118–119.

Bazyltchik S, Drozd VM, Reiners Chr, Gavrilin Yu (2001) Intellectual development of children exposed to radioactive iodine after the Chernobyl accident in utero and at the age under 1.5 years. In: Abstracts of the 3rd Int. Confer. «Health Effects of the Chernobyl Accident: Results of 15-year Follow up Studies», Kiev, June 4–8, 2001. *International Journal of Radiation Medicine*, Special Issue 3 (1–2):15.

Bebeshko V, Bazyka D, Nyagu A, Loganovsky K, Khomaziuk I, Lyashenko L, Klymenko V, Chumak A, Los I, Chumak V, Gaevaja L (2001) Radiation protection and health effects of Chernobyl nuclear power plant staff during the decommissioning. In: Abstracts of the 2nd International Conference «*The Effects of Low and Very Low Doses of Ionizing Radiation on Human Health*», 27–29 June, 2001, Dublin. World Council of Nuclear Workers, p. P6-6.

Biryukov A, Gorsky A, Ivanov S, Ivanov V, Maksioutov M, Meskikh N, Pitkevitch V, Rastopchin E, Souchkevitch G, Tsyb A (2001) Ed. by Souchkevitch GN, Repacholi MN: *Low Doses of Ionizing Radiation: Health Effects and*

Assessment of Radiation Risks for Emergency Workers of the Chernobyl Accident. Geneva: World Health Organization, 242 p.

Bomko MO (2004a): Morphometric neurovisual characteristic of organic brain damage in remote period of exposure to ionizing radiation as a result of the Chernobyl accident. *Ukrainian Medical Journal*, 2 (40): 96–101 [in Ukrainian].

Bomko MA (2004b) Morphometric neurovisual characteristic of organic brain damage in clean-up workers of the consequences of the Chernobyl accident in remote period of exposure to ionizing radiation. *International Journal of Psychophysiology*, 54 (1–2): 119.

Bomko MO (2005) *Structural-functional characteristic of organic mental disorders in clean-up workers of the consequences of the Chernobyl accident in remote period of exposure to ionizing radiation.* The dissertation for the academic degree of a Candidate of Medical Sciences in radiobiology. Research Centre for Radiation Medicine of Academy of Medical Sciences of Ukraine. *Kiev.*

Bromet EJ, Gluzman S, Schwartz JE, Goldgaber D. (2002) Somatic symptoms in women 11 years after the Chornobyl accident. *Environ Health Perspect, 110* (Suppl. 4), 625-629

Bromet EJ, Goldgaber D, Carlson G, Panina N, Golovakha E, Gluzman SF, Gilbert T, Gluzman D, Lyubsky S, Schwartz JE. (2000) Children's well-being 11 years after the Chornobyl catastrophe. *Arch Gen Psychiatry, 57(6):* 563-571

Buzunov VA, Strapko NP, Pirogova YeA, Krasnikova LI, Kartushin GI, Voychulene YuS, Domashevskaya TYe (2001) Epidemiology of non-cancer diseas es among Chernobyl accident recovery operation workers. *International Journal of Radiation Medicine*, 3 (3–4): 9–25.

Buzunov VA, supervisor (2003) *Pattern and Risks for Non-Cancer Chronic Diseases in Chernobyl Accident Survivors on the Base of Cohort Studies, Development of Models for Assessment and Forecasting taking into account Radiation and Non-Radiation Factors.* Final Report, Research Project 0100U003181, Kiev: Scientific Centre for Radiation Medicine of Academy of Medical Sciences [in Ukrainian].

Chuprikov AP, Pasechnik LI, Kryzhanovskaja LA, Kazakova SYe (1992) *Mental disorders at radiation brain damage.* Kiev Research Institute for General and Forensic Psychiatry [in Russian]

Danilov VM, Pozdeev VK (1994) The epileptiform reactions of the human brain to prolonged exposure to low-dose ionizing radiation. *Fiziol Zh Im I M Sechenova,* 80 (6): 88–98 [Article in Russian].

Demyttenaere K, Bruffaerts R, Posada-Villa J, Gasquet I, Kovess V, Lepine JP, Angermeyer MC, Bernert S, de Girolamo G, Morosini P, Polidori G, Kikkawa T, Kawakami N, Ono Y, Takeshima T, Uda H, Karam EG, Fayyad JA, Karam AN, Mneimneh ZN, Medina-Mora ME, Borges G, Lara C, de Graaf R, Ormel J, Gureje O, Shen Y, Huang Y, Zhang M, Alonso J, Haro JM, Vilagut G, Bromet EJ, Gluzman S, Webb C, Kessler RC, Merikangas KR, Anthony JC, Von Korff MR, Wang PS, Brugha TS, Aguilar-Gaxiola S, Lee S, Heeringa S, Pennell BE, Zaslavsky AM, Ustun TB, Chatterji S; WHO World Mental Health Survey Consortium (2004) Prevalence, severity, and unmet need for treatment of mental disorders in the World Health Organization World Mental Health Surveys. *JAMA* 291 (21): 2581–2590.

Ermolina LA, Sukhotina NK, Sosyukalo OD, Kashnikova AA, Tatarova IN (1996) The effects of low radiation doses on children's mental health (radiation-ontogenetic aspect). Report 2. *Social and Clinical Psychiatry*, 6 (3): 5–13 [in Russian].

European Commission. Radiation protection 100 (1998) *Guidance for protection of unborn children and infants irradiated due to parental medical exposures. Directorate-General Environment, Nuclear Safety, and Civil Protection* [http://www.europa.eu.int/comm/environment/radprot]

Fry RJ (2001) Deterministic effects. *Health Phys* 80 (4): 338–343.

Gamache GL, Levinson DM, Reeves DL, Bidyuk PI, Brantley KK (2005) Longitudinal neurocognitive assessments of Ukrainians exposed to ionizing radiation after the Chernobyl nuclear accident *Arch Clin Neuropsychol,* 20 (1): 81–93.

Gayduk FM, Igumnov SA, Shalckevich VB (1994) The complex estimation of neuro-psychic development of children undergone to radiation exposure in prenatal period as a result of Chernobyl disaster. *Social and Clinical Psychiatry,* 4 (1): 44–49 [in Russian]

Gelowitz DL, Rakic P, Goldman-Rakic PS, Selemon LD (2002) Craniofacial dysmorphogenesis in fetally irradiated nonhuman primates: implications for the neurodevelopmental hypothesis of schizophrenia. *Biological Psychiatry* 52 (7): 716–720.

Gourmelon P, Marquette C, Agay D, Mathieu J, Clarencon D. (2005) Involvement of the central nervous system in radiation-induced multi-organ dysfunction and/or failure. *BJR Suppl.* 27:62-68.

Gus'kova AK, Bisogolov GD (1971) *Radiation Sickness of Human.* Moscow: «Meditzina» Publishing House [in Russian].

Gus'kova AK, Shakirova IN (1989): Reaction of the nervous system on alterative ionizing irradiation. *Zh Nevrol Psikhiatr Im S S Korsakova* 89(2):138–142 [in Russian].

Gutin PH, Leibel SA, Sheline GE (Eds.) (1991): *Radiation injury to the nervous system.* New York: Raven Press.

Hall P, Adami HO, Trichopoulos D, Pedersen NL, Lagiou P, Ekbom A, Ingvar M, Lundell M, Granath F (2004) Effect of low doses of ionising radiation in infancy on cognitive function in adulthood: Swedish population based cohort study. *BMJ,* 328(7430): 19–24.

Havenaar JM, Cwikel JG, Bromet EJ. (eds.) *Toxic Turmoil: Psychological and Societal Consequences of Ecological Disasters.* New York, Kluwer Academic and Plenum Press, 2002

Havenaar JM, de Wilde EJ, van den Bout J, Drottz-Sjöberg B-M, van den Brink W. (2003). Perception of risk and subjective health among victims of the Chernobyl disaster. *Soc Sci Med,* 56:569-572

Havenaar JM, Rumyantseva GM, van den Brink W, Poelijoe NW, van den Bout J, van Engeland H, Koeter MWJ. (1997b) Long-term mental health effects of the Chernobyl disaster: an epidemiological survey in two former Soviet Regions. *Am J Psychiatry* 154:1605-1607

Havenaar JM, Rumyantzeva GM, Kasyanenko AP, Kaasjager K, Westermann AM, van den Brink W, van den Bout J, Savelkoul TJF. (1997a). Health effects of the Chernobyl disaster: illness or illness behaviour? A comparative general health survey in two former Soviet Regions. *Environ Health Perspect,* 105 (Suppl.6): 1533-1537

Honda S, Shibata Y, Mine M, Imamura Y, Tagawa M, Nakane Y, Tomonaga M (2002) Mental health conditions among atomic bomb survivors in Nagasaki. *Psychiatry and Clinical Neurosciences,* 56: 575–583.

Hunt WA (1987) Effects of ionizing radiation on behavior. In: Conklin JJ, Walker RI, editors. *Military radiobiology.* San Diego: Academic Press Inc. p. 321–330.

ICRP Publication 49 (1986) Developmental effects of irradiation on the brain of the embryo and fetus. A report of a Task Group of Committee 1 of the International Commission on Radiological Protection, 1986. In M.C. Thorne (Ed.). *Annals of the ICRP*, 16 (4). Oxford: Pergamon Press.

ICRP Publication 60 (1991) 1990 Recommendations of the International Commission on Radiological Protection. *Annals of the ICRP*, Vol. 21/1–3, Pergamon—Elsevier.

Igumnov S, Drozdovitch V. (2000). The intellectual development, mental and behavioral disorders in children from Belarus exposed in utero following the Chernobyl accident. *Eur Psychiatry,* 15:244-253

Igumnov SA (1996) Psychological development of children exposed to radiation in prenatal period as a result of Chernobyl disaster. *The Acta Medica Nagasakiensia.* 41 (3-4): 20–25.

Igumnov SA (1999) *The prospective investigation of a psychological development of children exposed to ionizing radiation in utero as a result of the Chernobyl accident.* The dissertation for the academic degree of a Doctor of Medical Sciences in radiobiology and psychiatry. State Scientific Center of Russian Federation Institute of Biophysics, Moscow.

Igumnov SA, Drozdovitch VV (2004) Antenatal exposure following the Chernobyl accident: neuropsychiatric aspects. *International Journal of Radiation Medicine,* 6 (1–4): 108–115.

Imamura Y, Nakane Y, Ohta Y, Kondo H (1999) Lifetime prevalence of schizophrenia among individuals prenatally exposed to atomic bomb radiation in Nagasaki City. *Acta Psychiatrica Scandinavia*, 100 (5), 344–349.

Ivanov VK, Maksioutov MA, Chekin Syu, Kruglova ZG, Petrov AV, Tsyb AF (2000) Radiation-epidemiological analysis of incidence of non-cancer diseases among the Chernobyl liquidators. *Health Physics* 78 (5): 495–501.

Ivanov VK, Maksiutov MA, Chekin SIu, Petrov AV, Tsyb AF, Biriukov AP, Kruglova ZG, Matiash VA (2005) The radiation risks of cerebrovascular diseases among the liquidators. *Radiats Biol Radioecol*, 45 (3): 261–270 [in Russian]

Jablensky A (2000) Epidemiology of schizophrenia: the global burden of disease and disability. *Eur Arch Psychiatry Clin Neurosci* 250 (6): 274–285.

Kesminiene AZ, Kurtinaitis J, Rimdeika G (1997) The study of Chernobyl clean-up workers from Lithuania. *Acta Med. Lituanica* 2: 55–61.

Kharchenko VP, Zubovskii GA, Kholodova NB (1995) Changes in the brain of persons who participated in the cleaning-up of the Chernobyl AES accident based on the data of radiodiagnosis (single-photon emission-computed radionuclide tomography, x-ray computed tomography and magnetic resonance tomography) *Vestn Rentgenol Radiol*, 1: 11–14 [in Russian].

Kholodova NB, Kuznetzova GD, Zubovsky GA, Kazakova PB, Buklina SB (1996) Remote consequences of radiation exposure upon the nervous system. *Zh Nevropatol Psikhiatr Im S S Korsakova,* 96 (5): 29–33.

Khomskaja ED (1995) Some results of neuropsychological study of Chernobyl accident consequences clean-up workers. *Social and Clinical Psychiatry,* 5 (4): 6–10 [in Russian].

Kimeldorf DJ, Hunt EL (1965) *Ionizing radiation: neural function and behavior.* New York, Academic Press.

Kolominsky Y, Igumnov S, Drozdovitch V (1999) The psychological development of children from Belarus exposed in the prenatal period to radiation from the Chernobyl Atomic Power Plant. *J Child Psychol Psychiatry*, 40 (2): 299–305.

Korr H, Thorsten Rohde H, Benders J, Dafotakis M, Grolms N, Schmitz C (2001) Neuron loss during early adulthood following prenatal low-dose X-irradiation in the mouse brain. *Int. J. Radiat. Biol.* 77 (5): 567–580.

Kovalenko AN, Loganovsky KN (2001): Whether Chronic Fatigue Syndrome and Metabolic Syndrome X in Chernobyl accident survivors are membrane pathology? *Ukrainian Medical Journal*, 6(26): 70–81 [in Russian].

Kozlova IA, Niagu AI, Korolev VD. (1999) The influence of radiation of the child mental development. *Zh Nevrol Psikhiatr Im S S Korsakova* 99(8):12–15 [Article in Russian].

Krasnov VN, Yurkin MM, Vojtsekh VF, Skavysh VA, Gorobets LN, Zubovsky GA, Smirnov YuN, Kholodova NB, Puchinskaja LM, Dudayeva KI (1993) Mental disorders in clean-up workers of the Chernobyl accident consequences. Report I: structure and current pathogenesis. *Social and Clinical Psychiatry,* 3 (1): 5–10 [in Russian]

Kusunoki Y, Kyoizumi S, Yamaoka M, Kasagi F, Kodama K, Seyama T (1999) Decreased proportion of CD4 T cells in the blood of atomic bomb survivors with myocardial infarction. *Radiat Res* 152 (5): 539–543.

Lebedinsky AV, Nakhilnitzkaja ZN (1960) *Ionizing radiation influence on the nervous system.* Moscow, Publishing House Atomizdat [in Russian].

Litcher L, Bromet EJ, Carlson G, Squires N, Goldgaber D, Panina N, Golovakha E, Gluzman S. (2000). School and neuropsychological performance of evacuated children in Kiev eleven years after the Chernobyl disaster. *J Child Psychol Psychiatry, 41*, 219-299

Loganovskaja TK (2004) Psychophysiological pattern of acute prenatal exposure to Ionizing radiation as a result of the Chernobyl accident. *International Journal of Psychophysiology*, 54 (1–2): 95–96.

Loganovskaja TK (2005) *Mental disorders in children exposed to prenatal irradiation as a result of the Chernobyl accident.* The dissertation for the academic degree of a Candidate of Medical Sciences in radiobiology. Scientific Centre for Radiation Medicine of Academy of Medical Sciences of Ukraine. Kiev.

Loganovskaja TK, Loganovsky KN (1999) EEG, cognitive and psychopathological abnormalities in children irradiated in utero. *Int J Psychophysiol* 34 (3): 213–224.

Loganovskaja TK, Nechayev SYu (2004): Psychophysiological effects in prenatally exposed children and adolescents as a result of the Chernobyl accident. *World of Medicine*, 4 (1): 130–137 [Article in Ukrainian]

Loganovsky K, Bomko M, Antypchuk Ye (2005b) Neuropsychiatric radiation effects: lessons from Chernobyl accident. Accepted Abstract *XIII World Congress of Psychiatry* Cairo, September 10-15, 2005 Egypt.

Loganovsky KN (1999) Clinical-Epidemiological aspects of psychiatric consequences of the Chernobyl disaster. *Social and Clinical Psychiatry*, 1(9):5–17 [in Russian].

Loganovsky KN (2000a) Neurological and psychopathological syndromes in the follow-up period after exposure to ionizing radiation. *Zh Nevrol Psikhiatr Im S S Korsakova.* 100(4):15-21 [in Russian]

Loganovsky KN (2000b) Vegetative-vascular dystonia and osteoalgetic syndrome or Chronic Fatigue Syndrome as a characteristic after-effect of radioecological disaster: the Chernobyl accident experience. *Journal of Chronic Fatigue Syndrome*, 7(3): 3–16.

Loganovsky KN (2002) *Mental disorders at exposure to ionising radiation as a result of the Chernobyl accident: neurophysiological mechanisms, unified clinical diagnostics, treatment.* The dissertation for the academic degree of a Doctor of Medical Sciences in radiobiology (03.00.01) and psychiatry (14.01.16). Scientific Centre for Radiation Medicine of Academy of Medical Sciences of Ukraine. Kiev.

Loganovsky KN (2003) Psychophysiological features of somatosensory disorders in victims of the Chernobyl accident. *Fiziol Cheloveka* 29(1): 122–130

Loganovsky KN, Bomko MA, Antipchuk YeYu, Denisyuk NV, Zdorenko LL, Rossokha AP, Chorny AI, Drozdova NV, Yukhimenko YeN, Kravchenko VI, Vasilenko ZL (2004b) Postradiation brain damage in remote period of the Chernobyl accident. *International Journal of Psychophysiology* 54 (1–2): 149.

Loganovsky KN, Bomko MO (2004) Structural-functional pattern of radiation brain damage in Chernobyl accident clean-up workers. *Ukrainian Medical Journal*, 5 (43): 67–74 [in Ukrainian].

Loganovsky KN, Kovalenko AN, Yuryev KL, Bomko MA, Antipchuk YeYu, Denisyuk NV, Zdorenko LL Rossokha AP, Chorny AB, Dubrovina GV (2003) Verification of organic brain damage in remote period of Acute Radiation Sickness. *Ukrainian Medical Journal*, 6(38): 70–78 [in Ukrainian].

Loganovsky KN, Loganovskaja TK. (2000) Schizophrenia spectrum disorders in persons exposed to ionizing radiation as a result of the Chernobyl accident. *Schizophr Bull* 26:751-773

Loganovsky KN, Nyagu AI, Loganovskaja TK (1999) Chronic fatigue syndrome — a possible effect of low and very low doses of ionizing radiation. In: Abstracts of the International Conference *«The effects of low and very low doses of ionizing radiation on human health»*, June 16–18, 1999, Versal, France. Versal: University of Versailles/St.Quentin-en-Yvelines, World Councol of Nuclear Workers (WONUC), p 14.

Loganovsky KN, Volovik SV, Flor-Henry P, Manton KG, Bazyka DA (2004a) Ionizing radiation as a risk factor for schizophrenia spectrum disorders. *International Journal of Psychophysiology*, 54 (1–2): 106.

Loganovsky KN, Volovik SV, Manton KG, Bazyka DA, Flor-Henry P (2005a) Whether ionizing radiation is a risk factor for schizophrenia spectrum disorders? *World Journal of Biological Psychiatry*, 6(4): (accepted)

Loganovsky KN, Yuryev KL (2001) EEG patterns in persons exposed to ionizing radiation as a result of the Chernobyl accident: part 1: conventional EEG analysis. *J Neuropsychiatry Clin Neurosci* 13 (4): 441–458.

Loganovsky KN, Yuryev KL (2004) EEG patterns in persons exposed to ionizing radiation as a result of the Chernobyl accident. Part 2: quantitative EEG analysis in patients who had acute radiation sickness. *J Neuropsychiatry Clin Neurosci* 16 (1): 70–82.

Loganovsky KN, Zdorenko LL (2004) Intelligence deterioration following acute radiation sickness. *International Journal of Psychophysiology*, 54 (1–2): 95.

Lysyanyj MI (1998) Current view on radiation effects on the nervous system. *Bulletin of Ukrainian Association of Neurosurgeons*, 7: 113–118 [in Ukraine].

McGale P, Darby SC. (2005) Low doses of ionizing radiation and circulatory diseases: a systematic review of the published epidemiological evidence. Radiat Res. 163(3): 247-57.

McGrath J., Saha S., Welham J., El Saadi Ossama, MacCauley C., Chant D. (2040) A systematic review of the incidence of schizophrenia: the distribution of rates and the influence of sex, urbanicity, migrant status and methodology. BMC Medicine, 2:13 http://www.biomedcentral.com/1741-7015/2/13

Mettler FA, Upton AC (Eds.) (1995) *Medical effects of ionizing radiation.* 2nd ed. Saunders W. B. Company, Philadelphia.

Mickley GA (1987) Psychological effects of nuclear warfare, in: Conklin JJ, Walker RI, editors. *Military radiobiology.* San Diego: Academic Press, Inc., pp 303–319.

Morozov AM, Kryzhanovskaja LA (1998) *Clinic, dynamic and treatment of borderline mental disorders in liquidators of Chernobyl accident.* Kiev, Publishing House Chernobylinterinform [in Russian].

Nakane Y, Ohta Y (1986) An example from the Japanese Register: some long-term consequences of the A-bomb for its survivors in Nagasaki. In: Ten Horn GHMM, Giel R, Gulbinat WH, Henderson JH (eds) *Psychiatric Case Registers in Public Health.* Elsevier Science Publishers B.V., Amsterdam, pp 26–27.

Napreyenko A, Loganovsky K (2001a) Psychiatric management of radioecological disaster victims and local wars veterans. *New Trends in Experimental and Clinical Psychiatry,* XVII (1–4): P. 43–48.

Napreyenko AK, Loganovskaja TK (2004) Mental disorders in prenatally irradiated children and adolescents as a result of the Chernobyl accident: diagnostic system, treatment and rehabilitation. *World of Medicine,* 4 (1): 120–129 [Article in Ukrainian].

Napreyenko AK, Loganovsky KN (1995) The systematics of mental disorders related to the sequelae of the accident at the Chernobyl Atomic Electric Power Station *Lik Sprava,* 5-6:25–29 [in Russian].

Napreyenko AK, Loganovsky KN (1997) *Ecological psychiatry,* Kiev Polygraphkniga Publishing House [in Russian].

Napreyenko AK, Loganovsky KN (1999) Current problems of emergency psychiatry at the radioecological disaster. In: *Emergency psychiatry in a changing world* Ed. by M. De Clercq, A. Andreoli, S. Lamarre, P. Forster. Amsterdam: Elsevier Science B.V., pp 199–202.

Napreyenko AK, Loganovsky KN (2001b) Ecological psychiatry. In: *Psychiatry,* edited by Napreyenko AK, *Kiev*: Zdorovja, Publishing House, pp 417–461 [in Ukrainian].

Niagu AI, Noshchenko AG, Loganovskii KN (1992) Late effects of psychogenic and radiation factors of the accident at the Chernobyl nuclear power plant on the functional state of human brain *Zh Nevropatol Psikhiatr Im S S Korsakova,* 92 (4): 72–77 [in Russian].

Nijenhuis MAJ, van Oostrom IEA, Sharshakova TM, Pauka HT, Havenaar JM, Bootsma PA. (1995). *Belarussian-Dutch Humanitarian Aid Project: "Gomel Project."* Bilthoven, National Institute for Public Health and Environmental Protection

Noshchenko AG, Loganovskii KN (1994) The functional brain characteristics of people working within the 30-kilometer area of the Chernobyl Atomic Electric Power Station from the viewpoint of age-related changes. *Lik Sprava,* 2: 16–19 [in Russian].

Nyagu A, Khalyavka I, Loganovsky K, Plachinda Yu, Yuriev K, Loganovskaja T. (1996c) Psychoneurological characterization of persons who had acute radiation sickness. *Problems of Chernobyl Excluzion Zone,* 3: 175–190 [in Russian].

Nyagu A, Loganovsky K, Loganovskaja T, Antipchuk Ye (1996a) The WHO Project on «Brain Damage in Utero»: mental health and psychophysiological status of the Ukrainian prenatally irradiated children as a result of the Chernobyl accident. In: *Bambino: Progetto Salute.* Proc. of Int. Meeting on prenatal, perinatal and neonatal neurological damages: preventive and therapeutic approach, Ancona, Italy, June 12-13, 1996, pp 34–58.

Nyagu AI, Cheban AK, Salamatov VA, Limanskaja GF, Yazchenko AG. Zvonaryeva GN, Yakimenko GD, Melina KV, Plachinda YuI, Chumak AA, Bazyka DA, Gulko GM, Chumak VV, Volodina IA (1993) Psychosomatic health of children irradiated in utero as a result of the Chernobyl accident. In: (Ed.) Nyagu AI. Proceedings of the International Conference «*Social, Psychological, and Pscyhoneurological Aspects of Chernobyl NPP Accident Consequences*», Kiev, September 28-30, 1992. Kiev: Association «Physicians of Chernobyl», Ukrainian Scientific Centre for Radiation Medicine, pp. 265–270 [in Russian]

Nyagu AI, Loganovsky KN (1998) *Neuropsychiatric effects of ionizing radiation*. Kiev, Publishing House Chernobylinterinform [in Russian].

Nyagu AI, Loganovsky KN, Cheban AK, Podkorytov VS, Plachinda YuI, Yuriev KL, Antipchuk YeYu, Loganovskaja TK (1996b) Mental health of prenatally irradiated children: a psychophysiologycal study. *Social and Clinical Psychiatry* 6 (1): 23–36 [Article in Russian].

Nyagu AI, Loganovsky KN, Chuprovskaya NYu, Kostychenko VG, Vaschenko EA, Yuryev KL, Zazymko RN, Loganovskaya TK, Myschznchuk NS (2003): Nervous system. In: Vozianov A, Bebeshko V, Bazyka D, editors. *Health Effects of Chornobyl Accident. Kiev*: DIA, pp 143–176.

Nyagu AI, Loganovsky KN, Loganovskaja TK, Repin VS, Nechaev SYu (2002a) Intelligence and brain damage in children acutely irradiated in utero as a result of the Chernobyl accident. In: Imanaka T, editor. *KURRI-KR-79. Recent Research Activities about the Chernobyl NPP Accident in Belarus, Ukraine and Russia*. Kyoto: Research Reactor Institute, Kyoto University, pp 202–230.

Nyagu AI, Loganovsky KN, Loganovskaja TK. (1998). Psychophysiologic aftereffects of prenatal irradiation. *Int J Psychophysiol,* 30, 303-311

Nyagu AI, Loganovsky KN, Nechayev SYu, Repin VS, Antipchuk YeYu, Loganovskaya TK, Bomko MA, Yuryev KL, Petrova IV, Pott-Born R (2004a) Potential effects of prenatal brain irradiation as a result of the Chernobyl accident in Ukraine. In: *Abstracts of International Workshop on «The French-German Initiative Results and their Implication for Man and Environment»*, Kiev, October 5–6, 2004, p. 38.

Nyagu AI, Loganovsky KN, Pott-Born R, Nechayev SYu, Repin VS, Antipchuk YeYu, Loganovskaya TK, Bomko MA, Yuryev KL, Petrova IV (2004b) Franco-German Initiative for Chernobyl, Project N°3 «Health Effects on the Chernobyl Accident», Sub-Project No 3.4.1: *«Potential Effects of Prenatal Irradiation on the Brain as a Result of the Chernobyl Accident (Ukraine)»*, Final Report.

Nyagu AI, Loganovsky KN, Pott-Born R, Repin VS, Nechayev SYu, Antipchuk YeYu, Bomko MA, Yuryev KL, Petrova IV (2004c) Effects of prenatal brain irradiation as a result of the Chernobyl accident. *International Journal of Radiation Medicine*, 6 (1–4): 91–107.

Nyagu AI, Loganovsky KN, Yuryev KL (2002b) Psychological consequences of nuclear and radiological accidents: delayed neuropsychiatric effects of the acute radiation sickness following Chernobyl. In: *Follow-up of delayed health consequences of acute accidental radiation exposure. Lessons to be learned from their medical management.* IAEA-TECDOC-1300, IAEA, WHO. Vienna: IAEA, P. 27–47

Nyagu AI, Loganovsky KN, Yuryev KL, Zdorenko LL (1999) Psychophysiological aftermath of irradiation. *International Journal of Radiation Medicine* 2 (2): 3–24.

Otake M, Schull WJ (1984) In utero exposure to A-bomb radiation and mental retardation: A reassessment. *British Journal of Radiology* 57: 409–414.

Otake M, Schull WJ (1998) Radiation-related brain damage and growth retardation among the prenatally exposed atomic bomb survivors. *International Journal of Radiation Biology*, 74 (2), 159–171.

Otake M, Schull WJ, Lee S (1996) Threshold for radiation-related severe mental retardation in prenatally exposed A-bomb survivors: a re-analysis. *International Journal of Radiation Biology*, 70 (6), 755–763.

Preston DL, Shimizu Y, Pierce DA, Suyama A, Mabuchi K (2003) Studies of mortality of atomic bomb survivors. Report 13: Solid cancer and noncancer disease mortality: 1950–1997. *Radiat Res* 160(4):381–407.

Rahu M, Tekkel M, Veidebaum T, Pukkala T, Hakulinen A, Auvinen A, Rytomaa T, Inskip PD, Boice JD Jr. (1997). The Estonian study of Chernobyl clean-up workers: II. Incidence of cancer and mortality. *Radiation Res, 147,* 653-657

Rajendran R, Raju GK, Nair SM, Balasubramanian G (1992) Prevalence of oral submucous fibrosis in the high natural radiation belt of Kerala, south India. *Bull World Health Organ.* 1992;70(6):783-9.

Revenok AA (1998) Psychopathic-like disorders in persons with an organic brain lesion as a result of exposure to ionizing radiation *Lik Sprava*. 3: 21–24 [in Russian].

Revenok OA (1999) *Structural and dynamical characterization of organic brain damage in persons exposed to ionizing radiation as a result of the Chernobyl accident* The dissertation for the academic degree of a Doctor of Medical Sciences in psychiatry. Ukrainian Research Institute for General and Forensic Psychiatry, Kiev.

Romanenko AY, Nyagu AI, Loganovsky KN, Tirmarche M., Gagniere B., Buzunov VA, Ledoschuk BA, Trocyuk N, Yuryev KL, Zdorenko LL, Antipchuk YeYu, Bomko MA, Loganovskaya TK, Petrova IV, Kartushin GN, Khmelko VYe, Paniotto VI, Zakhozha VA (2004a) Franco-German Initiative for Chernobyl Project N°3 «Health Effects on the Chernobyl Accident» Sub-Project No 3.4.8: «*Data Base on Psychological Disorders in the Ukrainian Liquidators of the Chernobyl Accident*», Final Report.

Romanenko AY, Nyagu AI, Loganovsky KN, Tirmarche M., Gagniere B., Buzunov VA, Ledoschuk BA, Trocyuk N, Kartushin GN, Khmelko VYe, Paniotto VI, Zakhozha VA, Yuryev KL, Zdorenko LL, Antipchuk YeYu, Bomko MA, Loganovskaya TK, Petrova IV (2004b) Data base on psychological disorders in the Ukrainian liquidators of the Chernobyl accident. In: Abstracts of International Workshop on «*The French-German Initiative: Results and Their Implication for Man and Environment*», October 5-6, 2004, Kiev, pp 56–57.

Romodanov AP, Vynnyts'kyj OR (1993) Brain lesions in mild radiation sickness. *Lik Sprava,* 1: 10–16 [in Ukrainian].

Ron E, Modan B, Flora S, Harkedar I, Gurewitz R (1982) Mental function following scalp irradiation during childhood. *Am. J. Epidemiol.* 116: 149–160.

Rumyantseva GM, Chinkina OV, Levina TM, Margolina VYa (1998) Mental disadaptation of the Chernobyl accident clean-up workers. *Russian Medical Journal. Contemporary Psychiatry,* 1 (1): 56–63 (http://www.rmj.ru /sovpsih/t1/n1/8.htm; http://www.rmj.ru/p1998_01/8.htm)

Sasaki H, Wong FL, Yamada M, Kodama K (2002) The effects of aging and radiation exposure on blood pressure levels of atomic bomb survivors. *J Clin Epidemiol* 55(10):974–981.

Schindler MK, Wang L, Selemon LD, Goldman-Rakic PS, Rakic P, Csernansky JG (2002) Abnormalities of thalamic volume and shape detected in fetally irradiated rhesus monkeys with high dimensional brain mapping. *Biological Psychiatry* 51 (10): 827–837.

Schull W, Otake M Yoshimaru H (1988) Effect on intelligence test score of prenatal exposure to ionising radiation in Hiroshima and Nagasaki: A comparison of the T65DR and DS86 dosimetry systems. *Technical Report RERF 3–88*. Hiroshima: Radiation Effects Research Foundation.

Schull WJ (1997) Brain damage among individuals exposed prenatally to ionizing radiation: a 1993 review. *Stem Cells*, 15 Suppl 2: 129–133.

Schull WJ, Otake M (1999) Cognitive function and prenatal exposure to ionizing radiation. *Teratology*, 59 (4), 222–226.

Shabadash AL (1964) Cytochemical characterization of reactivity and inhibit-protective states of the nervous cells at radiation injuries. In: *Restoration processes at radiation injuries*. Moscow: Atomizdat Publishing House, pp. 53–60 [in Russian].

Shimizu Y, Pierce DA, Preston DL, Mabuchi K (1999) Studies of the mortality of atomic bomb survivors. Report 12, part II. Noncancer mortality: 1950–1990. *Radiat Res* 152(4):374–389.

Torubarov FS, Blagoveshchenskaia VV, Chesalin PV, Nikolaev MK (1989) Status of the nervous system in victims of the accident at the Chernobyl atomic power plant. *Zh Nevrol Psikhiatr Im S S Korsakova* 89 (2): 138–142 [in Russian].

Trocherie S, Court L, Gourmelon P, Mestries J-C, Fatome M, Pasquier C, Jammet H, Gongora H, Doloy MTh (1984) The value of EEG signal processing in the assessment of the dose of gamma or neutron-gamma radiation absorbed dose. In: Court L, Trocherie S, Doucet J (eds) *Le Traitment du Signal en Electrophysiologie Experimentale et Clinique du Systeme Nerveux Central*, Vol. II, pp. 633–644.

Tsirkin SIu (1987) International study of schizophrenia based on a WHO program. Comparative characteristics of the data. *Zh Nevropatol Psikhiatr Im S S Korsakova* 87 (8): 1198–1203 [in Russian]

Turuspekova ST (2002) Neuropsychological functions in individuals exposed to small dose ionizing radiation. *Zh Nevrol Psikhiatr Im S S Korsakova* 102 (3): 16–19 [in Russian].

UNSCEAR 1982 Report to the General Assembly, with Scientific Annexes. United Nations Scientific Committee on the Effects of Atomic Radiation. United Nations, New York, Vol. 2, Biological Effects.

UNSCEAR 1993 Report to the General Assembly (1993): Annex I. Late deterministic effects in children. Sources and effects of ionizing radiation. *UNSCEAR 1993 Report to the General Assembly, with Scientific Annexes*. United Nations Scientific Committee on the Effects of Atomic Radiation. United Nations, New York, pp. 899–908.

UNSCEAR 2000 report to the General Assembly (2000), Annex J. Exposures and effects of the Chernobyl accident. *International Journal of Radiation Medicine*, Special issue, 2–4 (6–8): 3–109.

UNSCEAR 2000 report to the General Assembly, with scientific annexes. Sources and effects of ionizing radiation. United Nations Scientific Committee on the Effects of Atomic Radiation. New York: United Nations, Vol. I: Sources.

Vasculescu T, Pasculescu G, Papilian V, Serban I, Rusu M (1973) The effect of low X-ray doses on the central nervous system. *Radiobiol. Radiother. (Berl.)*, 14 (4): 407–416.

Viinamäki H, Kumpusalo E, Myllykangas M, Salomaa S, Kumpusalo L, Kolmakov S, Ilchenko I, Zhukowsky G, Nissinen A. (1995). The Chernobyl accident and mental wellbeing - a population study. *Acta Psychiatr Scand,* 91: 396-401

Voloshina NP (1997) *Structural and functional brain disorders in patients with dementia of different genesis.* The dissertation for the academic degree of a Doctor of Medical Sciences in neurology and psychiatry. Kharkiv Institute of Advanced Medical Studies, Kharkiv, Ukriane.

Volovik S, Loganovsky K, Bazyka D (2005) Chronic Fatigue Syndrome: molecular neuropsychiatric projections. Abstract *XIII World Congress of Psychiatry* Cairo, September 10-15, 2005 Egypt, p. 225.

Vyatleva OA, Katargina TA, Puchinskaya LM, Yurkin MM (1997) Electrophysiological characterization of the functional state of the brain in mental disturbances in workers involved in the clean-up following the Chernobyl atomic energy station accident. *Neurosci Behav Physiol,* 27 (2): 166–172.

World Health Organization (1996) *Health consequences of the Chernobyl accident. Results of the IPHECA pilot projects and related national programmes.* Geneva, World Health Organization

World Health Organization (2001) *World Health Report 2001. Mental health: New understanding, new hope.* Geneva: World Health Organization.

World Health Organization (2005) Health Effects of the Chernobyl Accident and Special Health Care Programmes. Report of the UN Chernobyl Forum Expert Group «Health» (EGH), Working Draft, August 31, 2005, 179 p. http://www.iaea.org/NewsCenter/Focus/Chernobyl/index.shtml

Yaar I, Ron E, Modan B, Rinott Y, Yaar M, Modan M (1982) Long-lasting cerebral functional changes following moderate dose X-radiation treatment to the scalp in childhood: an electroencephalographic power spectral study. *J. Neurol. Neurosurg. Psychiatry* 45 (2): 166–169.

Yaar I, Ron E, Modan M, Perets H, Modan B (1980) Long–term cerebral effects of small doses of X–irradiation in childhood as manifested in adult visual evoked responses. *Ann. Neurol.* 8: 261–268.

Yamada M, Izumi S (2002) Psychiatric sequelae in atomic bomb survivors in Hiroshima and Nagasaki two decades after the explosions. *Soc. Psychiatry Psychiatr. Epidemiol.* 37: 409–415

Yamada M, Kasagi F, Sasaki H, Masunari N, Mimori Y, Suzuki G (2003) Association between dementia and midlife risk factors: the radiation effects research foundation adult health study. *J Am Geriatr Soc* 51(3):410–414.

Yamada M, Kodama K, Wong FL. (1991). The long-term psychological sequelae of atomic bomb survivors in Hiroshima and Nagasaki. In Ricks R., Berger M.E., O'Hara R.M. (Eds), *The medical basis for radiation preparedness III: The psychological perspective.* New York: Elsevier, pp. 155-163

Yamada M, Sasaki H, Mimori Y, Kasagi F, Sudoh S, Ikeda J, Hosoda Y, Nakamura S, Kodama K (1999) Prevalence and risks of dementia in the Japanese population: RERF's adult health study Hiroshima subjects. Radiation Effects Research Foundation. *J Am Geriatr Soc* 47(2):189–195.

Yamada M, Wong FL, Fujiwara S, Akahoshi M, Suzuki G (2004) Noncancer disease incidence in atomic bomb survivors, 1958–1998. Radiat. Res., 161: 622–632.

Zdorenko LL, Loganovsky KN (2002) Mental working capacity in liquidators with organic mental disorders in the remote period after the Chernobyl accident. *Ukrainian Medical Journal*, 4 (30): 120–126 [in Ukrainian].

Zhavoronkova LA, Gabova AV, Kuznetsova GD, Sel'skii AG, Pasechnik VI, Kholodova NB, Ianovich AV (2003) Post-radiation effect on the interhemispheric asymmetry in EEG and thermography characteristics. *Zh Vyssh Nerv Deiat Im I P Pavlova* 53 (4): 410–419 [in Russian].

Zhavoronkova LA, Gogitidze NV, Kholodova NB (1996) The characteristics of the late reaction of the human brain to radiation exposure: the EEG and neuropsychological study (the sequelae of the accident at the Chernobyl Atomic Electric Power Station) *Zh Vyssh Nerv Deiat Im I P Pavlova* 46 (4): 699–711 [in Russian]

Zhavoronkova LA, Gogitidze NV, Kholodova NB (2000) Postradiation changes in the brain asymmetry and higher mental functions of right- and left-handed subjects (the sequelae of the accident at the Chernobyl Atomic Electric Power Station) *Zh Vyssh Nerv Deiat Im I P Pavlova* 50 (1): 68–79 [in Russian].

Zhavoronkova LA, Kholodova NB, Gogitidze NV, Koptelov IuM (1998) A dynamic assessment of the reaction of the human brain to radiation exposure (the aftermath of the accident at the Chernobyl Atomic Electric Power Station). *Zh Vyssh Nerv Deiat Im I P Pavlova* 48 (4): 731–742 [Article in Russian].

Zhavoronkova LA, Kholodova NB, Zubovskii GA, Smirnov IuN, Koptelov IuM, Ryzhov NI (1994) The electroencephalographic correlates of neurological disorders in the late periods of exposure to ionizing radiation (the aftereffects of the accident at the Chernobyl Atomic Electric Power Station) *Zh Vyssh Nerv Deiat Im I P Pavlova* 44 (2): 229–238 [in Russian].

Zhavoronkova LA, Kholodova NB, Zubovskii GA, Smirnov YuN, Koptelov YuM, Ryzhov NI (1995a): Electroencephalographic correlates of neurological disturbances at remote periods of the effect of ionizing radiation (sequelae of the Chernobyl' NPP accident). *Neurosci Behav Physiol* 25 (2): 142–149.

Zhavoronkova LA, Kholodova NB, Zubovsky GA, Gogitidze NV, Koptelov YM (1995b) EEG power mapping, dipole source and coherence analysis in Chernobyl patients. *Brain Topogr*, 8 (2): 161–168.

Zozulya YuP (ed.) (1998) *Chronic influence of small doses of ionizing radiation: experimental studies and clinical observations.* Kiev, Publishing House Chernobylinterinform [in Ukrainian].

CHAPTER 4

The Influence of the Chernobyl Accident on Wild Vertebrate Animals.

Dr. Eugene Yu. Krysanov, PhD MD,
Institute of Ecology and Evolution
Russian Academy of Sciences
Moscow
Russia

In this chapter I review briefly the biological effects in natural populations of vertebrate animals exposed to the radiation from the Chernobyl accident.

For radiological effects, the most important stage was the initial period after the accident. In this period radioactive nuclides accumulated in different systematic and ecological groups of animals and, accordingly, these systems suffered the highest dose load and radiation effects. The ecological consequences have not stabilized. On radioactively polluted land several processes are occurring.

From the point of revealing the consequences of radio-ecological exposures, there are a number of biocenoses which were initially contaminated. First of all there is the region of the "Red Forest" and the cooling tank of Chernobyl NPS. On the territory of this "Red Forest" there was a sharp decrease in the population of murine rodents. This resulted from the effects of high dose loads; later on, these populations were gradually recreated (Taskaev *et al.*, 1988).

In small mammals, studies showed increased embryonic death and an increase in the level of variable cells of the dominant lethal mutations, of reciprocal translocations, deviations of the morphological content of blood and embryonic pathologies (Zainullin, 1998; Pomerantzeva, Ramaja, 1999; Sokolov *et al.*, 1999).

In the cooling pond of Chernobyl NPS the death of separate hydrocoles was not found. There was, however, a short-term depression of benthic organisms. In fish – the most sensitive hydrocoles – there were changes on organ indices, on tissue and on cytogenic levels. It was reported that, in 31 species of fish, the maximum influence of ^{137}Cs occurred in predator species of the lithophilous group (spawning on stony soil). These include zander and asp. In 1998 the dose rate received by fish, by embryos and by larvae from the bottom sediments was about 10-20 cGy/day, from the water – 2-3 mGy/day. During 1989-1990, the survival of silver carp was studied from the moment of spawning and the dose of radiation of both individuals was 7 - 8 Gy. Biochemical analysis showed that the energy sources, the reproductive capacity and general indices did not differ from the norm, but 8 % of males were sterile. The young fish of silver carp received 15 cGy. The Cytogenetic analysis revealed 22,7% of deviations in the irradiated fish (5-7 % in the control group), but it was noted that in the pond there was not only the radiation factor, but chemical and thermal influences also (Ryabov 2004).

In other territories, where the radiation levels were not so acute, there was no mass death of animals, no cases of expressed organ radiation disease and no radio-specific anomalies in the breeding or in the behavior of animals. In discussing the high radio-resistance of natural eco-systems in general and in its specific elements in animals, it is of interest to mention the numerous radiation effects that occur in lower levels of biological organization. In this connection the ecological indexes, which are used for the characteristics of populations and of animal communities, seem to show the influence of only large doses of radiation. The influence of small doses of radiation seems to be apparent only at the molecular and cellular levels. The numerous data on the radiation effects in different species of animals that have been reported at levels lower than the eco-system level of biological organization and on territories situated far enough from the most radioactive areas are the proof of this.

Among such effects on wild animals that inhabit radiation polluted biocenoses we may include the cytogenic deviations in amphibians and in small mammals. There have been reports of differences in many levels of biochemical parameters between mammals in the polluted zone and those in clean territories. Destructive changes in liver and in Hemopoietic organs, an increase in the response of the immune system, changes in energy, carbohydrate and in albumin kinetics and in their regulation, genetic effects and the increase of fluctuating asymmetry have all been reported by many researchers observing murine rodents and also in some other species of wild animals in the 30-km zone relative to regions with lower levels of pollution. (Biological ..., 1990; Zakharov & Krysanov, 1996).

In addition, it was reported that there were defects in the process of growth and in the development of sex cells and their structure in fish (oocyte and nucleus measurement aberrations, aberrations in oocyte development, asyncronous oocyte growth, thickening of follicular membrane, nucleus disintegration, etc.). The authors concluded that there was a negative influence in the effects of continuous radiation exposure of both the environment and the tissues and organs in fish gametogenesis. (Petukhov & Kokhnenko, 1998)

Low radiation doses (absorbed dosage rate from internal and external irradiation of order of 0.4–5.5 m Gy/day) were shown to induce negative biological effects in pond carp reared in radiocontaminated ponds (specific activity of bottom sediments — 801–3235 Bq/kg). In particular, it was reported that there were reductions in reproductive parameters in stripped fishes, a rise in the frequency of morphological changes and cytogenetic injuries in embryos and larvae. A rise in the frequency of morphological changes and cytogenetic injuries in carp aged 1-summer was also noted (Slukvin & Goncharova, 1999).

Cytomorphological analysis of pike gonads was conducted in another study. This revealed some deviations from the normal development of oocytes. A major part of oocytes at the period of vitellogenesis in fish from lakes Perstok and Smerzhov resulted from germ cells with degenerative changes. The thickness of the radial membrane in pike egg cells reached 25- 30 mcm; its normal thickness being about 10 mcm. Deviations in female gonad development and in gonads of pike were also observed. It was concluded that a negative influence of reservoir radiocontamination on reproductive process could, in future, result in a reduction in the reproductive ability of the above-studied species populations (Kokhenko, 2000).

One study compared the results obtained from *Apodemus agrarius*, *Microtus oeconomus* and *Microtus arvalis* caught in the spring of 1987, which inhabited territories with low ((0,02 - 0,1 mR/h), medium (2 - 20 mR/h) and high (150 - 200 mR/h) levels of gamma-radiation. Results revealed several kinds of hepatocyte dystrophy, deviations in the energy systems efficiency in the organ. Decrease in the activity of SDG, PDG and LDG were the most essential changes, but there was no significant connection with the level of external radiation (Shishkina *et al.*, 1992).

Another study reported a statistically significant increase in the level of cytogenetic injuries (aberrant cells) in bone marrow and the intestinal epithelium of amphibians and rodents, in the alveolar macrophages of rodents, as well as an increase in the micronuclei level in erythrocytes of amphibian peripheral blood. (Yeliseeva *et al.*, 1996).

Goncharova *et al.* (1996) reported a high frequency of chromosome aberrations and a rise in the level of genome mutations up to 1991 (decreasing with the dose absorbed). They noted an indication of increased sensitivity of hereditary structures in the somatic cells of new animal generations, descendants of the generations of 1986-1988 and, hence, argued that this showed an absence of genetic adaptation to the mutagenic effect of radio-contamination in natural populations during the whole period of their investigations.

The detected cytogenetic effects of chronic low-intensive irradiation in the germ and somatic cells of wild animals exceeded the expected levels deduced from extrapolation of the data from the high-dose range of acute or chronic irradiation. In wild murine rodents increased frequencies of cytogenetic injuries in somatic and germ cells, as well as embryonal lethality, were shown to remain over the life spans of no less than 22 generations (Goncharova & Ryabokon, 1998).

Another study shows that there appears to be a malfunction of erythropoiesis function at low radiation exposure. After 20 generations of field voles from contaminated areas, an excess level of mutation process is taking place in somatic cells, despite an essential decrease of radiation exposure (Smolich & Ryabokon, 1997).

As usual, registered changes are not linearly connected with the power of the absorbed dose. This underlines the difficulty of predicting radiation effects, especially at the low dose level.

Similarly, there seems to be no direct dependence between the level of the accumulation of the radioactive nuclides and the biological effects (the frequencies of the variable cells, micronuclei, and fluctuating asymmetry).

The increasing radio-resistance, with the passage of time, of populations existing in polluted biogeocenoses has, however, been reported at low levels of pollution. (Il'enko & Krapivko, 1994).

Most of the research had been carried out on small mammals with a lesser number of studies devoted to this problem focusing on other groups of animals (fish and amphibians).

In general, the majority of studies are devoted to the problems of accumulation and transformation of radioactive nuclides by animals (Ryabov, 2004; Lebedeva & Ryabtzev, 1999; Bondarkov *et al.*, 2002, 2002a; Kuchmel et al, 2003; Oleksik *et al.*, 2002;).

It is worth noting that most studies were carried out within the first decade after the accident rather than the second.

Thus, the long term biological effects in natural populations of animals, long after the initial damage, are less likely to have been studied well. Some of these studies show long-term effects.

One interesting study examined genetic diversity in bank voles from Chernobyl-contaminated sites and uncontaminated sites. The workers sampled three geographic sites; Oranoe, a reference site with virtually no radioactive contamination (<2 Ci/km2) and two highly contaminated sites; Glyboke Lake and the "Red Forest" (both 1,000 Ci/km2). Genetic diversity in the population from the "Red Forest" (0.722 ± 0.024) was significantly greater than at the Oranoe reference site (0.615 ± 0.039), while genetic diversity at Glyboke Lake (0.677 ± 0.068) was intermediate (Matson *et al.*, 2000).

The level of erythroblasts with micronuclei in peripheral blood in *Rana temporaria* larvae caught in the stream in the Central Botanical Gardens area (National Academy of Sciences of Belarus, Minsk) was 6 times higher than the control — the population from Berezinsky reserve (p<0.001). A significant excess (3- 4 times) of the micronuclei number as against the control (p<0.001) was observed in brown frog larvae caught near v. Veprin in the Chericov district of the Mogilev region. The number of erythrocytes with micronuclei in all the populations of brown frogs from radiocontaminated areas caught before 1991 was shown to be higher than in the control and, in some cases, by a factor of 30 (p<0.001). During long-term monitoring of the populations of both brown and narrow-muzzled frogs inhabiting radio-contaminated areas it was discovered that, besides an increased yield of cytogenetic damage in bone marrow cells and erythrocytes, different levels of change took place in the ratio of erythrocyte number in peripheral blood. Deviations in erythron towards an increase in the specific weight of small and large cells were noted in 1988 - 1989 in animals of all the populations inhabiting radiocontaminated areas (Voitovich, 2000).

According to our own data, the frequency of aberrant cells in some species of small mammals remains rather high even today. Among bank vole they are about 6-8% in the most polluted areas of the Bryansk Region. These values are higher then the control values by 3-4 times. In addition, in the populations of red field-voles we studied, a high frequency of individuals had maximum values of aberrant cells (2005 data).

High values of aberrant cells in some species of hydrocoles from the tanks of 10 km zone of Chernobyl accident were also noted. The frequency of aberrant cells in perch from the cooling pond of Chernobyl NPP is about 6%, which is 2-times higher than the background values. In Crucian carp, from Glubokoe Lake, the frequency of variable cells is about 10 % (Gudkov *et al.*, 2004).

The frequency of aberrant cells in the lake frog from lake Glubokoe is about 6-7% (author data 2005), which is lower then noted in 1988, but 2-3 times higher then the spontaneous (Krysanov, 1993).

At the same time, it was found that there was a decrease in the level of reciprocal translocations and of micronucleus in natural populations of small mammals 15-18 years after the damage. (Pomerantzeva *et al.*, in press, Rodgers & Baker. 2000).

From this short review we may conclude that, with regard to populations of wild animals inhabiting radioactively contaminated territories, the situation has altered from its pre-contamination one and continues to change today.

It seems that significant but subtle damage caused the Chernobyl NPP radiation is a factor which has caused different and sometimes unobvious changes in these populations, which, as a consequence, may lead to the transformation of the karyotype of the gene pool and to other transformations also.

References

Biological and radioecological aspects consequences accident on Chernobyl NPP. 1990. Moscow. 230p.

Bondarkov M.D., Gashak S.P., Goryanaya Ju. A., Maksimenko A.M., Ryabushkin A.N., Salyi O.V., Shulga A.A., Awan S., Chesser B.E., Rodgers B.E.,Some characteristics of 90Sr and 137Cs metabolism in newborn bank voles Int. Cher. Center, Kiev, 2002, 477-485.

Bondarkov M.D., Gashak S.P., Goryanaya Ju. A., Maksimenko A.M., Shulga A.A., Chigevsky I.V., Oleksik T.K.,Features of radioactive contamination of amphibians of Chernobyl region. 2002. Int. Cher. Center, Kiev, 508-517

Goncharova R.I., Ryabokon N.I. The results of long-term monitoring of animal populations chronically irradiated in radiocontaminated areas. Radiological catastrophe in Chernobyl: report of international study. Tokyo, 1998. P. 256–266. (Jap)

Goncharova R.I., Ryabokon N.I., Slukvin A.M.. Dynamics of mutability in somatic and germ cells of animals inhabiting regions of radioactive fall-out. Cytology and Genetics. 1996. V. 30, № 4. P. 35-41. (Rus)

Gudkov D.I., Derevets V.V., Kuzmenko M.I., Nazarov A.B., Krot Yu.G., Kipnis L.S., Mardarevich M.G., Syvak O.V.Hydrobionts of the Chernobyl NPP exclusion zone: present-day level of radioactive contamination dose rates and cytogenetic effects. 2004., Proc. II Int. Conf., Tomsk, 167-171. (Rus).

Il'enko A.I., Krapivko T.P., Radioresistance of bank voles populations (Clethrionomys glareolus) in conditions of radioactive contamination. Proc. Natl. Acad. Sci. Rus., 1994. v. 336, N5, 714-718. (Rus)

Krysanov E. Yu. 1993. Chromosome aberrations in lake frogs from Chernobyl zone in 1987. In: Impact of Radiation from Chernobyl accident. Proc. First All Union Conf., "Gidrometeoisdat", St-Pet., v. 2, 27-29 (In Russian)

Kuchmel S.V., Deryabina T.G., Ostudin I.A., Antonuk G.A., Specific activity of 137Cs in organs and tissues of wild hoofed of Polessye reservation. 6th Ann. Conf. Int. Cern. Center, 2003, 216 (Rus).

Lebedeva N. V., Ryabtzev I.A. Accumulation of radionuclides in birds. 1999. Bioindication of radioactive pollutions Moscow. Nauka, 72-85 (Rus)

Matson C.W., Rodgers B. E., Chesser R.K.,Baker R.J. Genetic diversity of Clethrionomys glareolus populations from highly contaminated sites in the Chornobyl region, Ukraine. 2000, Envir. Toxicol. Chem., 19, 2130-2135

Oleksik T. R., Gashak S.P., Glenn T.S., Jagoe C.H., Peles J. D., Purdue J. R., Tsyusko O.V., Zalissky O.O., Smith M.H., Frequency distributions of 137Cs in fish and mammal populations. J. Envir. Rad., 2002, v. 61, 55-74.

Petukhov V.B., Kokhnenko O.S. Gametogenesis of bream and roach under radioactive contamination of cisterns of Belarus. Proc. Natl. Acad Sci. Bel., Ser. Biology. 1998. N 3. P. 115-120. (In Russian)

Pomerantzeva M D., Ramaya L.K., 1999. Genetical effects of high level of radiation in mice on contaminated area of Chernobyl (in Bioindication of radioactive contamination (Ed. D. Krivolutzky), 187-199. (In Russian)

Pomerantzeva M D., Ramaya L.K., Rubanovich A.N. 2006. Genetical consequences of the increased radiating background at rodents. Radiobiology and Radioecology (in press).

Rodgers B. E. and R. J. Baker. 2000. Frequencies of micronuclei in bank voles from zones of high radiation at Chernobyl, Ukraine. Environmental Toxicology and Chemistry. 19:1644-1649.

Ryabov I.I., Radioecology of fishes of reservoirs in region influences of accident on the Chernobyl atomic power station, 2004, Moscow, KMK, 215 c. (In Russian)

Shishkina L. N., Matteriy L.D., Kudyasheva A.G., Zagorskaya N.G., Taskaev A.I. Structurally functional infringements in a liver of wild mammals from areas of accident}on the Chernobyl atomic power station. Radiobiology. 1992. v. 32, N 1, 19-29. (Rus)

Slukvin A.M., Goncharova R.I. Radiation-induced biological effects in pond carp and measures of their prevention. Proc. of Intern. Conf. On Present State and Prospects of Aquaculture Development, Gorki, 7 – 9 December 1999. Gorki, 1999. P.136–138. (Rus)

Smolich I.I., Ryabokon N.I. Micronucleus frequency in somatic cells of red vole (clethorionomus glareolus) from the populations of chronic radiation exposure. Proc. Natl. Acad Sci. Bel., Ser. Biology. 1997. № 4. P. 42-46. (Rus)

Sokolov V.E., Krylova T.E., Skurat L.N., 1999. Embryonic development of rodent in chronic irradiatin impact on forest biogeocenoses (in Bioindication of radioactive contamination (Ed. D. Krivolutzky), 123-128 (In Russian)

Sushchenya L.M., Pikulik M.M., Plenin A.E. (eds), 1995. Animals in the zone of the Chernobyl atomic power electric station. Minsk "Navuka i tekhnika", 263 p. (In Russian)

Taskaev A.I., Testov B.V., Matteriy L. D., 1988, Ecological and morphoecological consequences accident on the Chernobyl NPP for populations of small mammals, Syktyvkar, 56. (In Russian)

Testov V.V., Taskaev A.I., Ryabov I.N., Ryabtsev LA. 1993. Changes in abundance of rodents on the site with various pollution levels. In: Impact of Radiation from Chernobyl accident. Proceedings of the First All Union Conference. (Obninsk, June 1988), vol. 2, Ed. Yu.A. Israel, St-Pet.., "Gidrometeoisdat", 147-150. (In Russian)

Voitovich A.M. Micronuclei frequency in erythrocytes and disturbance in differentiation process in brown-frog erythron under chronic radiation exposure // Proc. Natl. Acad. Sci., 2000. N.3. P.60- 63. (rus).

Yeliseeva K.G., Kartel N.A., Voitovich A.M., Trusova V.D., Ogurtsova S.E., Krupnova E.V. Chromosome aberrations in various tissues of murine rodents and amhpibians. Cytology and Genetics. 1996. V. 30, № 4. P. 20-24. (Rus)

Zainullin Genetical effect of chronic irradiatin in low doses of ionizing radiation, 1998. St. Petersburg, Nauka, 99p. (In Russian)

Zakharov V.M. & Krysanov E.Yu., Eds., 1996 Consequences of the Chernobyl Catastrophe: Environmental Health, Moscow, 160. (In Russian)

CHAPTER 5

Chromosome Aberrations in the Blood Lymphocytes of People Exposed to Radiation as a Result of the Chernobyl Accident.

G.P.Snigiryova[1], V.A.Shevchenko[2]

[1]*Federal State Institution Russian Scientific Center of Roentgenology & Radiology Roszdrav, Moscow, Russia*
[2]*Vavilov Institute of General Genetics, Moscow, Russia*

In the process of examining people exposed to radiation, a large volume of information has been accumulated during the twenty years since the accident at the Chernobyl nuclear power plant. The negative effects of the accident adversely affected the health of many people. Among them are the liquidators– the largest and most seriously affected category of irradiated population of our country and the residents of settlements within the zones of radioactive fallout.

The most important role of these examinations is the prediction of genetic effects of ionizing radiation on the human organism and the assessment of the risk of future development of pathologies. These tasks can be successfully accomplished only with reliable information on accumulated irradiation doses. Unfortunately, in an emergency situation such as took place at the Chernobyl NPP in 1986, it is very difficult to assess the real radiation load with only the help of the limited physical dosimetry data available at that time. The biological methods, namely, the cytogenetic ones, based on the analysis of the frequency of chromosome aberrations make a considerable contribution to solving these problems. These methods allow a quantitative estimation of the effect of radiation on the human organism taking into account its individual peculiarities (first of all, individual radio-sensitivity) and the condition of the organism at the time of irradiation. The data obtained with these methods provide a more accurate prediction of possible early and remote consequences of irradiation.

The principles of the cytogenetic method of dosimetry were convincingly substantiated in numerous works in Russia and abroad and the results of those works served as the basis for the recommendations worked out by WHO, IAEA and UNSCEAR on the practical use of chromosomal analysis of blood lymphocytes in determining the doses of exposure.

The radiobiological basis for employing cytogenetic methods in radiation investigations is a high sensitivity of blood lymphocytes and the availability of radiation-specific chromosome rearrangements, the dependence of which on dose has been studied for most types of ionizing radiation.

The action of radiation leads to the occurrence of chromosome aberrations of the exchange type - dicentrics, polycentrics, centric rings, translocations and acentric fragments. Of primary interest in studying these radiation effects, especially in the course of quantitative analysis, is the frequency of dicentrics and translocations. The analysis of dicentrics is successfully applied in biodosimetric estimations of the doses of irradiation shortly (within a year) after a single, relatively uniform, radiation exposure. The advantage of the analysis of dicentrics is their easy detection with a light microscope without the complicated procedures of treatment and staining of chromosome preparations. The limited application of this method is due to the fact that cells with such aberrations, being genetically unbalanced, may be eliminated in the course of the cell cycle. As the postradiation period increases, the uncertainty in dose estimations on the basis of the frequency of dicentrics also significantly increases. In this case the residual frequency of dicentrics in peripheral blood lymphocytes permits the presence or absence of radiation damage to be established.

A more promising method in biological dosimetry is the analysis of translocations. Using this method it is possible to determine the degree of radiation injury in the organism and to estimate the irradiation dose within a long period after the exposure. This is possible due to the fact that, in contrast to dicentrics, translocations do not disturb the process of mitosis and therefore can easily be transmitted through a series of cell generations, i.e., the frequency of translocations does not significantly change with the time after irradiation

The spontaneous frequency of dicentrics and translocations is characterized by relatively low values and averages between 0 and 5 per 10,000 cells for dicentrics. The spontaneous frequency of translocations is approximately one order of magnitude higher. Using the frequency of chromosome aberrations it is possible to estimate relatively low doses. The lower limit of sensitivity is 0.01 Gy for dicentrics and about 0.25 Gy for translocations [1].

Despite the restrictions of the analysis of dicentrics used for retrospective estimation of irradiation doses, it has found wide use in cytogenetic monitoring in groups of people exposed to irradiation in an emergency. An increased level of chromosome aberrations in blood lymphocytes may precede the development of pathological processes in the human organism and therefore information on the extent of genome damage is extremely important in the formation of groups at risk of developing various diseases, including oncological ones [2,3].

Cytogenetic examinations acquired special importance in connection with the accident at the Chernobyl NPP, due to which large contingents of people were exposed to irradiation of varying intensity. Much attention in literature was initially given to the cytogenetic examination of the liquidators of the accident and populations from regions with radioactive contamination. During the examination of 158 workers of the nuclear power plant and liquidators at the clinic of the Institute of Biophysics, irradiation doses in the range from 0.1 to 13.7 Gy were estimated by means of cytogenetic analysis [4]. The accuracy of these estimates was confirmed by the later clinical state of the patients.

In the course of cytogenetic monitoring of the liquidators Maznik et al. [5] established that the average dose calculated by the data of cytogenetic analysis was higher than the values officially recorded in the documents.

Pilinskaya et al. [6] noted that during examination of the liquidators with a known irradiation dose, the positive correlation between the cytogenetic results and the data of physical dosimetry was not always observed. The dose values calculated on the basis of cytogenetic analysis were, as a rule, lower, which, in the opinion of the authors, is due to the elimination of unstable chromosome aberrations with the post-irradiation time.

However, one cannot but note that there are numerous works [7,8,9,10,11] in which evidence for an increased level of unstable chromosome aberrations (dicentrics, centric rings and acentric fragments) in blood lymphocytes of the liquidators is presented. According to the authors, the observed changes are retained over many years after the radiation exposure. Sevan'kaev et al. [8] carried out a cytogenetic examination of 875 liquidators (1-6 years after the accident) who worked in Chernobyl for two weeks to three months. An increased frequency of the indicators of radiation damage – dicentrics and centric rings – was observed only in groups of liquidators who worked between 1986-1988. In those who worked in 1989 the frequency of chromosome aberrations did not differ from the control values. Estimation of individual doses was not possible because of too low values of the cytogenetic indices. Nevertheless, the average doses calculated for the examined groups by the frequency of dicentrics were in agreement with the data of personal dosimeters. Bogomazova [9] carried out a cytogenetic examination of the population irradiated as a result of the accident at the Chernobyl NPP (liquidators and people evacuated from regions contaminated with radionuclides). 4-10 years after the accident the frequency of unstable chromosome aberrations in the blood of the examined people exceeded the control level.

Much attention in the literature was given to the cytogenetic examination of children living on radioactive territories.

In the course of 17 years, Sevan'kaev *et al.* [10] carried out cytogenetic examinations of children and teenagers from contaminated territories of the Kaluga region. Throughout the whole period of examination an increased level of unstable chromosome aberrations was observed in 30-60% of the examined subjects. The authors did not discover an increase in the frequency of chromosome aberrations with an increase in the dose load.

While examining children living in radionuclide-contaminated areas of the Zhitomir and Kiev regions, Pilinskaya *et al.* [11] established that the highest level of chromosome aberrations was in children living within a territory with a maximum level of exposure.

It should be noted that some ambiguity in the data presented in the works of various researchers is due to an insufficient number of examined cells (sometimes no more than 100 cells), which, in some cases, may lead to incorrect conclusions.

In the early 1990s there appeared works in which the FISH method was used for analyzing stable chromosome aberrations in people who suffered from the Chernobyl accident.

The results of examination of the liquidators with large doses of irradiation are presented in the work of Sevan'kaev *et al.* [12]. Retrospective estimation of irradiation doses by the frequency of dicentrics and translocations was made 5 years after the accident. The dose values obtained by the two methods were in agreement.

Data for an increase in the frequency of stable translocations in blood lymphocytes of people exposed at different times to irradiation of different character are presented in the work [13].

On the basis of the data obtained in the course of analysis of the frequency of stable translocations in blood lymphocytes of the liquidators, determination of irradiation doses was made 5-10 years after the Chernobyl accident [14]. The average dose for the examined group of liquidators was 9 cGy (from 0 to 51 cGy). These values were lower compared to the data of physical dosimetry. Little correlation (<0.2) was noted between the biological and physical doses.

This work presents the results of cytogenetic examination of people exposed to irradiation as a result of the accident at the Chernobyl NPP. The possibility of using cytogenetic data for assessing the doses of irradiation and predicting the negative impacts of radiation exposure is also considered.

Results and Discussion

Cytogenetic examination of the liquidators. Biodosimetry.

During their activities in Chernobyl the liquidators were exposed to short-term, prolonged or fractional irradiation with different dose rates. Partial and rather incomplete information on irradiation doses determined with the physical methods of dosimetry does not permit an objective assessment of the radiation damage. Taking into consideration that the level of chromosome rearrangements in blood lymphocytes correlates with the value of irradiation dose, a more accurate estimate of the consequences of the accident in terms of both individual and population dosimetry can be obtained through cytogenetic analysis.

In 1986 443 liquidators were cytogenetically examined using the classical method (analysis of dicentrics). Blood samples were collected a few days after the end of their work within the 30 km zone. Table 1 shows the data of the cytogenetic analysis and biological dosimetry for the examined group of liquidators. As follows from the presented data, the frequency of dicentrics is 16.5 times higher than the control level. The irradiation doses were estimated using a calibration dose-effect curve for dicentrics [15] constructed on the basis of experimentally obtained data (acute gamma-irradiation of blood samples at a dose rate of 0.103 Gy/min). The curve is described by a linear quadratic equation:

$$Y = (0.10 \pm 0.03) + (1.5 \pm 0.4)*D + (6.3 \pm 0.3)*D^2,$$

where Y is the frequency of dicentrics per 100 cells, D is the dose of irradiation, Gy.

Table 1. Frequency of chromosome aberrations (classical method) in a group of dators examined in 1986.

Groups	Number of patients	Number of cells	*dic ± m/ 100 cells	Biological estimate of dose, Gy (95% C.I.)
Liquidators	443	41927	0.33 ± 0.01*	0.16 (0.07 – 0.42)
Control	114	51430	0.02 ± 0.01	

significance of differences as compared to the control, p<0.05; dic = dicentrics

The average dose for the examined group of liquidators was 0.16 Gy. According to the data [16], an average dose of irradiation for a representative group of liquidators from Russia constituted 0.12 Gy. The average values of irradiation doses for the liquidators working in Chernobyl in 1986 approached 0.16 Gy [17].

In the period from 1992 to 1995 a cytogenetic analysis with the use of the FISH method was carried out for 52 liquidators who participated in the restoration works in Chernobyl since 1986. The results of this analysis and biological dosimetry for the given group are presented in Table 2.

Table 2. The results of cytogenetic examination (FISH method) and average irradiation doses in the liquidators

Groups	Number of patients	Number of cells	Translocations $F_G \pm m$/ 100 cells	Biological estimate of dose, Gy (95%)
All liquidators	52	44283	1.20 ± 0.16*	0.27(0.05-0.50)
Liquidators - 1986	35	27767	0.86 ± 0.13*	0.19(0.05-0.50)
Liquidators - 1986-1995	17	15516	1.81 ± 0.35*	0.39(0.10-0.80)
Control	15	21953	0.47 ± 0.09	-

* significance of differences as compared to the control, p<0.05*

The frequency of translocations in the examined group exceeds 2.5-fold the control level. 17 examined liquidators worked in Chernobyl periodically from 1986 to 1995, i.e., they were exposed to long-term fractional irradiation. The frequency of translocations in this group is 4 times higher than in the control group and 2 times higher than in the group of liquidators (35 patients) who worked in Chernobyl only in 1986. This may be due to the fact that the liquidators who repeatedly took part in the clean-up works over several years received a higher total dose compared to the liquidators who worked in Chernobyl only in 1986. The irradiation doses were calculated by the frequency of translocations using a calibration dose-effect curve [18]. An average dose for the whole examined group of liquidators (52 persons) constituted 0.27 Gy. For the group of liquidators who repeatedly worked in Chernobyl in the period from 1986 to 1995 the dose of irradiation made up 0.39 Gy and for the liquidators who worked only in 1986 - 0.19 Gy.

It should be noted that the doses estimated by the frequency of translocations and dicentrics for the liquidators of 1986 are in good agreement with and do not significantly contradict the data obtained with the physical methods of dosimetry.

The doses determined for the liquidators by means of cytogenetic analysis are equivalent to the doses of acute single irradiation as they were calculated with the help of calibration curves plotted for acute irradiation with a rather high dose rate (0.5 Gy/min). In reality, the liquidators were, as a rule, exposed to long-term fractional low-intensity irradiation. Therefore, in some cases, the real irradiation doses might be much higher. This largely concerns the liquidators who came to work in Chernobyl several times over a long period. This important fact must be taken into account when analyzing the health condition of the liquidators.

Cytogenetic monitoring of the liquidators.

Figure 1 presents the results of long-term analysis of the frequency of chromosome aberrations (namely, dicentrics) in blood lymphocytes of the liquidators who took part in the restoration works in Chernobyl in 1986. In the first year of examination the average frequency of dicentrics in the group of liquidators was 0.33 per 100 cells and exceeded the control level 16.5-fold. In succeeding years the value of this parameter gradually decreased (approximately 2-fold) compared to the first year of examination. However, throughout the whole period of examination, up to 2004, the frequency of dicentrics in blood lymphocytes of the liquidators remained significantly higher compared to the control values – from 0.10 to 0.14 per 100 cells.

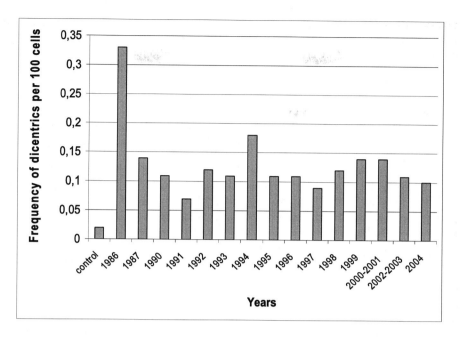

Fig.1 Average frequency of dicentrics in a group of liquidators examined at different times after the accident

The fact of an increase in the frequency of cytogenetic disturbances in the liquidators was confirmed by numerous investigations and is now beyond question. However, the interpretation of the results involves certain difficulties. There is an opinion that genome destabilization is the basic pathogenetic mechanism for the development of various somatic diseases. Despite the lack of direct evidence for the influence of increased chromosome variability in somatic cells on the health of the liquidators, a considerable amount of recent data demonstrates the contribution of somatic mutations to the development of a number of diseases. At present, the results of investigations concerning somatic pathologies and general morbidity among liquidators remain rather contradictory. Many researchers revealed unfavorable tendencies in the dynamics of some classes of somatic diseases in the liquidators: an increase in the incidence of diseases of the cardiovascular and osteoarticular systems, nervous system, psychical sphere, as well as disturbances in metabolism and cases of immunodeficiency. Examinations of the patients' dynamics reveal progressive deterioration in the state of their health [19]. Functional disturbances are transformed with time into organic ones and a mild form of a disease into a grave one.

Cytogenetic monitoring of the liquidators in combination with the examination of their health condition will permit reliable information to be gathered on the correlation between the frequency of chromosome aberrations and the risk of development of somatic diseases.. At present there are literature data [20, 21] on a significant correlation between the frequency of chromosome aberrations and the risk of carcinogenesis. Therefore it seems to be important during cytogenetic examination of the liquidators to differentiate groups with different genetic risks depending on the level of chromosome aberrations in peripheral blood lymphocytes. Patients with a high degree of genetic risk need constant medical examination for the negative consequences of irradiation to be established and prevented.

Cytogenetic examination of the Bryansk region residents.

Between 1992-1994 cytogenetic examinations were carried out among residents of the Bryansk region living in areas contaminated with radionuclides after the accident at the Chernobyl nuclear power plant. It should be emphasized that, due to differences in the levels of contamination and in the radionuclide composition of the radioactive fallout and also due to a wide range of biogeochemical characteristics in the territory, the cytogenetic examination of the population for the purpose of assessing the injurious action of radiation acquires particular importance. Under such circumstances the data obtained with only the help of the physical methods of dosimetry cannot be used for unbiased assessment of the negative impacts of irradiation.

Table 3 presents the results of examinations of the population in the Bryansk region. Since no significant differences were revealed in the cytogenetic parameters between the groups living in the zones of different social-economic status, all examined people were united in one group. The results of cytogenetic analysis were compared with the control data to draw a conclusion about the possible harmful action of radiation on the human cell genome. In the group of examined residents of the Bryansk region the total frequency of chromosome aberrations was 1.43 per 100 cells. This value is 2 times higher than in the control group. Ionizing radiation induces mainly aberrations of the chromosomal type, first of all dicentrics, in cells of living organisms. The frequency of dicentrics in the examined group made up 0.1 per 100 cells and exceeded 5-fold that in the control group. These data are in good agreement with the results of cytogenetic examination of children and teenagers from the Kaluga region living on radionuclide-contaminated territories [10].

Table 3. The results of cytogenetic examination (classical method) of the population of the Bryansk region

	No. of patients	No. of cells	ab± m/ 100 cells	(dic+ Rc) ± m/ 100 cells	ace ± m/ 100 cells	chr ± m/ 100 cells
Bryansk region	80	21027	1.43 ± 0.08*	0.10 ± 0.02*	0.36 ± 0.04	0.93 ± 0.07
Control	114	51430	0.66 ± 0.04	0.02 ± 0.01	0.23 ± 0.02	0.41 ± 0.03

significance of differences compared to the control, p<0.05
ab – total number of aberrations;   dic – dicentrics; R_c - centric rings;   ace – acentric fragments;   chr – aberrations of the chromatid type.

An increased level of dicentrics - indicators of radiation damage – in people from contaminated territories suggests a constant mutagenic influence of the radiation factors on the human organism.

It should also be noted that the level of chromatid aberrations in the examined group is 2 times higher compared to the control value. Chromosome aberrations of this type occur from the action of mutagens predominantly in S- and G_2-phases of the cell cycle, which is characteristic for an overwhelming majority of chemical agents and viruses [22]. The increased level of chromatid aberrations may be due to chemical environmental pollution both as a result of the accident and as a result of intensive application of pesticides and chemical fertilizers. One cannot rule out the influence on the cell's genetic apparatus of constant long-term low-intensity irradiation from radionuclides entering the human organism from the environment with food products.

Conclusion

To sum up the results of long-term monitoring of cytogenetic effects in the Chernobyl liquidators and in people from contaminated territories it is necessary to state the following;

- analysis of chromosome aberrations in blood lymphocytes is one of the most important and necessary tasks when examining the victims of the Chernobyl accident;
- the results of cytogenetic analysis (classical and FISH) can be successfully used for indication and quantitative assessment of the effects of radiation on the human organism. The doses calculated on the basis of cytogenetic examination take into account such important factors as individual radiosensitivity and the condition of health at the time of irradiation. Hence the obtained information permits a more exact prediction of unfavorable consequences of irradiation, including the estimation of genetic risk;
- analysis of morbidity in the groups of irradiated people examined with cytogenetic methods will help determine the correlation between the frequency of chromosome aberrations in somatic cells and the risk of development of pathological processes. This will be of use in working out sensitive criteria for the early detection of negative changes in the health condition of exposed people.

The authors are sincerely grateful to the scientific workers of the Cytogenetic Laboratory of RSC RR N.N.Novitskaya, E.D.Khazins and A.N.Bogomazova for technical assistance in preparing this publication.

References

1. United Nations. Sources and Effects of Ionizing Radiation. UNSCEAR 2000 Report to the General Assembly with Scientific Annexes. United Nations sales publication. No.E.00.IX.4. New York. 2000.
2. Snigiryova G.P., Lyubchenko P.N., Shevchenko V.A. et al. // Hematol. and Transfusiol. 1994. V.39. N 3. P.19-21. (in Russ.)
3. Kholodova N.B., Zubovsky G.A., Snigiryova G.P. IIIrd scientific-practical regional conference on "The Condition of Health of the Liquidators of the Accident at the Chernobyl NPP at a Distant Period". Moscow. 2004. P.176. (in Russ.)
4. Retrospective dosimetry of liquidators of the Consequences of the Accident at the Chernobyl Nuclear Power Plant. By ed.: Kruychkov V.P. and Nosovskii A.V., Kiev, 1996, P.256. (in Russ.)
5. Maznik N.A., Vinnikov V.A., Toyplaya V.A. et al. Proceedings of International Conference "Emergency Situations: Notification and Liquidation of Consequences", Kharkov, 2000, P.200-205. (in Russ.)
6. Pilinskaya M.A., Shemetun A.M., Bondar' A.Yu., Dyibskii S.S. // Vestnik AMS SSSR. 1991. N 8. P.40-45. (in Russ.)
7. Pilinskaya M.A., Shemetun A.M., Bondar' A.Yu., Dyibskii S.S. // Cytol. And Genetics. 1996. N 2. P.17-25. (in Russ.)

8. Sevan'kaev A.V. // Rdaiat. Biology. Radioekology. 2001. V.40. N5. P.589-595 (in Russ.)

9. Bogomazova A.N. The study of stable and unstable chromosome aberrations in people exposed as a results of the Chernobyl accident in late term after irradiation. Master's thesis. S.-Petersburg. 2000. (in Russ.)

10. Sevan'kaev A.V., Mikhailova G.F., Potetnya O.I. // Radiat. Biology. Radioecology. 2005. V.45. N 1. P.5-15.

11. Pilinskaya M.A., Shemetun A.M., Dyibskii S.S. et al. // Cytol. and Genetics. 1994. N3. P.27-32. (in Russ.)

12. Sevan'kaev A.V., Lloyd D.C., Edwards A.A. and Moiseenko V.V. // Radiation Protection Dosimetry. 1995. V.59. N2. P.85-91. (in Russ.)

13. Vorobtsova I.E., Mikhel'son V.M., Vorob'eva M.V. et al. // Radiat. Biology. Radioecology. 1994. V.34. N6. P.798-803. (in Russ.)

14. Moor D.H., Tucker J.D., Jones I.M. et al. // Radiation Research. 1997. V.148. P.463-475.

15. Snigiryova G.P., Bogomazova A.N., Novitskaya N.N. et al. // Proc. Intern. Conference "Genetic Consequences of Emergency Radiation Situations". Moscow: RUDN. 2002. P. 313 - 328.

16. Waison A.A., Zhakov I.G., Knizhnikov V.A. et al..// Med. Radiology. 1990. N10.
P. 21-28. (in Russ.)

17. Liberman A.N. Radiation and reproductive health. S.- Petersburg: 2003. 225 p. (in Russ.)

18. Bauchinger M., Schmid E., Zitzelsberger H. et al. // Int. J. Radiat. Biol. 1993. V. 64.
P. 179-184.

19. Kharchenko V.P., Kholodova N.B., Snigiryova G.P. // Medical Anthropology. Minsk: 2004. P.213-214. (in Russ.)

20. Bonassi S.,Znaor A., Norppa H., Hagmar L.. // Cytogenet. Genome Res.2004. V. 104. P.376-382.

21. Rossner P., Boffetts P., Ceppi M. et al. // Environ. Health Perspectives. 2005. V.113. N5. P.517-520.

22. Natarajan T., Boei J., Darroudi F. et al // Environ. Health Perspectives. 1996. V.104.
N 3. P. 445-448.

CHAPTER 6
Radiation- induced Effects in Humans After *in utero* Exposure: Conclusions from Findings After the Chernobyl Accident

Short title: Teratogenic Effects after Chernobyl

Inge Schmitz-Feuerhake*
Department of Physics, University of Bremen, Germany

Abstract. A threshold dose of 100 mSv is assumed for radiation-induced teratogenic effects in publication No. 90 of the ICRP (2003), which is based on the findings in the Japanese A-bomb survivors. A variety of observations about congenital malformations, foetal loss, stillbirths, and infant deaths, as well as Down´s syndrome after the Chernobyl release make manifestly clear the incompleteness and inadequacy of the Japanese data in these areas. The radiation-induced developmental effects of Chernobyl, especially at great distances, are usually denied because of the low values of the estimated human exposures. Biological dosimetry in contaminated regions, however, shows that physical dose estimations lead to considerable underestimations of the true exposure. The assumptions about teratogenic effects by incorporated radionuclides have to be revised.

Introduction

The evaluation of radiaton risks by international radiation protection committees is based on findings concerning Japanese A-bomb survivors. The effects observed there after prenatal exposure were mental retardation and reduced head size, while no other significant detriment was detected. The period between the 8th and 15th week of gestation is thought to be the period of greatest risk.

As was pointed out by different, researchers the Japanese data suffer, however, from several restrictions which limit their suitability as a general base for deriving radiation risks. One point is a probable and proven severe selection bias because of the catastrophic situation after the bombing. Another objection which must be stressed, especially considering perinatal effects, is the fact that the investigations of the Radiation Effects Research Foundation (RERF) in Hiroshima did not begin earlier than 5 years after the catastrophe when the RERF research institute was established there. The completeness of the data must therefore be put into question.

The reduced selection of assumed effects and the assertion of a threshold dose as high as 100 mSv by the ICRP contradicts former findings in experimental research (Table 1) and several observations in humans previous to the Chernobyl accident. Tables 2-5 list the results of studies in regions affected by fallout from the latter.

Table 1. Minimum dose below 100 mSv which showed significant effects after *in utero* x-ray exposure in experimental studies (data taken from Fritz-Niggli, 1997)

	Dose mSv	Days after conception	Effects	Reference
Mice	10 50 50 50	8 0.5 0.5; 1.5 7.5	Cumulated developmental defects Death of the embryo Death of the embryo, polydactyly Death of the embryo, skeletal malformations	Michel, Fritz-Niggli 1977 Rugh, Grupp 1959 Ohzu 1965 Jacobsen 1966
Rats	10 50	18 0.4; 0.7	Reflex distortions Foetal death	UNSCEAR 1986 Roux et al. 1983

Table 2 Observed increase of congenital malformations *in utero* after exposure by the Chernobyl accident

Country	Effects	References
Belarus National Genetic Monitoring Registry	Anencephaly, spina bifida, cleft lip and/or palate, polydactyly, limb reduction defects, esophageal atresia, anorectal atresia, multiple malformations	Lazjuk et al. 1997
Belarus Highly exposed region of Gomel	Congenital malformations	Bogdanovich 1997; Savchenko 1995
Chechersky district of the Gomel region	Congenital malformations	Kulakov et al. 1993
Mogilev region	Congenital malformations	Petrova et al. 1997
Brest region	Congenital malformations	Shidlovskii 1992
Ukraine Polessky district of the Kiev region	Congenital malformations	Kulakov et al. 1993
Lygyny region		Godlevsky, Nasvit 1998
Turkey	Anencephaly, spina bifida	Akar et al.1988/89; Caglayan et al. 1990; Güvenc et al. 1993; Mocan et al. 1990

Bulgaria, region of Pleven	Malformations of heart and central nervous system, multiple malformations	Moumdjiev et al. 1992
Croatia	Malformations by autopsy of stillborns and cases of early death	Kruslin et al. 1998
Germany German Democratic Republic, Central registry	Cleft lip and/or palate	Zieglowski, Hemprich 1999
Bavaria	Cleft lip and/or palate Congenital malformations	Scherb, Weigelt 2004 Korblein 2003a, 2004; Scherb, Weigelt 2003 Strahlentelex 1989
Annual Health Report of West Berlin 1987	Malformations in stillborns	Lotz et al. 1996
City of Jena (Registry of congenital malformations)	Isolated malformations	

Table 3 Observed increase of stillbirths, infant deaths, spontaneous abortions and low birth weight after in utero exposure by the Chernobyl accident

Country	Effects	References
Belarus Selected regions	Perinatal deaths*)	Petrova et al. 1997
Chechersky District near Gomel	Perinatal deaths	Kulakov et al. 1993
Gomel region	Perinatal deaths	Korblein 2003a,b
Ukraine Polessky District near Kiev	Perinatal deaths, reduced birth rate**), premature births	Kulakov et al. 1993
Lygny region	Early neonatal deaths	Godlevsky, Nasvit 1998 Korblein 2003a,b
Zhitomir oblast, Kiev region, Kiev City	Perinatal deaths, reduced birth rate	
Europe: Greece, Hungary, Poland,	Stillbirths	Scherb et al. 1999b, 2000b, 2003

Sweden	Infant mortality	Korblein 2003a
Poland	Spontaneous abortions	Ulstein et al. 1990
Norway	Low birth weight	Czeisel 1988
Hungary	Premature births among malformed children	Harjulehto et al. 1989
Finland	Reduced birth rate Stillbirths	Harjulehto et al. 1991 Scherb, Weigelt 2003
Germany Total (FRG + GDR)	Perinatal deaths	Korblein, Kuchenhoff 1997; Scherb et al. 2000a,2003
Southern Germany	Early neonatal deaths	Lüning et al. 1989
Bavaria	Perinatal deaths, stillbirths	Grosche et al. 1997; Scherb et al. 1999a, 2000a, 2003
	Reduced birth rate	Korblein 2003a

*) Perinatal deaths summarize stillbirths and deaths in the first 7 days from birth
**) Reduced birth rate is considered as a measure for spontaneous abortions

Table 4 Increase of Down´s syndrome after *in utero* exposure by the Chernobyl accident

Region	**Results**	**References**
Belarus National Genetic Monitoring Registry	Excess 1987-1994 ca. 17 %	Lazjuk et al. 1997
Western Europe	Beginning 1 yr after the accident, reaching 22% within 3 years	Dolk et al. 1999
Sweden	"Slight" excess in most exposed areas (30 %)	Ericson, Kallen 1994
Scotland, Lothian region (0.74 million inhabitants)	Excess peak in 1987 (2-fold significant)	Ramsay et al. 1991
South Germany	Investigations of amniotic fluid	Sperling et al. 1991
Berlin West	Sharp increase after 9 months	Sperling et al. 1991, 1994

Table 5 Observed health defects in children after *in utero* exposure by the Chernobyl accident except malformations and Downs´s syndrome

Belarus Selected regions	Mental disorders Speech-language disorders, mental retardation	Kondrashenko et al. 1996 Kolominsky et al. 1999 Kulakov et al. 1993
Chechersky District near Gomel	Diseases of respiratory organs, blood, circulation etc.	Sychik, Stozharov 1999a,b
Stolin District in Brest region	Diseases of respiratory organs, glands, blood, circulation, digestive organs	
Belarus, Ukraine, Russia	Mental retardation and other mental disorders	Kozlova et al. 1999
Ukraine Polessky District near Kiev	Diseases of respiratory organs, blood, circulation etc.	Kulakov et al. 1993 Ponomarenko et al. 1993
Rovno Province	Childhood morbidity	
Immigrants to Israel from contaminated areas	Asthma	Kordysh et al. 1995

Estimated exposures from the Chernobyl accident

In spite of the overwhelming evidence of effects observed also in countries far away from the Chernobyl event they are ignored by international radiation protection committees. They assume that the exposures of the population are much too low to generate teratogenic effects. Indeed, the physical dose estimates resulted in mean effective life-time exposures in large regions of Europe and in Turkey below 1.2 mSv (UNSCEAR 1988). The highest average dose for a sub-region in the first year after the accident is derived to 2 mSv in Belarus.

These estimates show, however, large discrepancies in comparison with biological dosimetry. Investigations of unstable and stable chromosome aberrations in the lymphocytes of persons in the contaminated regions have been carried out by a variety of research groups in rather large collectives directly after the accident or some years later. Dicentric chromosomes can be considered as radiation-specific and most sensitive because of their very low and nearly constant rate in unexposed persons (Hoffmann and Schmitz-Feuerhake, 1999).

It is generally found that the observed rates of dicentric chromosomes after Chernobyl are considerably higher – by 1 or 2 orders of magnitude – than would be expected from physically derived dose estimates. This evaluation is possible although the dose-effect relationships in cases of incorporated radioactivity are not known. For low dose homogenous exposure by low LET irradiation the rate of dicentrics can be considered as dose-proportional which follows from studies in the range of background exposure. In European countries, far from the Chernobyl site, the exposure of the tissues, except the thyroid, is assumed to be mainly generated by ^{137}Cs and ^{134}Cs which distribute homogenously inside the body. The whole body doubling dose for dicentrics by homogenous low LET radiation is about 10 mSv (Hoffmann and Schmitz-Feuerhake, 1999). Elevations of dicentrics in persons which are higher than 2-fold would therefore mean that the whole body dose exceeds 10 mSv. Such elevations have been found manifold after Chernobyl. Findings for Austria, Germany and Norway are shown in Table 6.

A remarkable finding in many of the chromosome studies is that they report an over-dispersion of the dicentrics and the occurrence of multi-aberrant cells (Bochkov and Katosova 1994; Hille et al. 1995; Salomaa et al. 1997; Scheid et al. 1993, Sevan`kaev et al. 1993; Stephan and Oestreicher 1989; Verschaeve et al. 1993). This is an indication of a relevant contribution of incorporated α-activity which is not considered adequately in the physical dose estimates.

Table 6 Chromosome aberrations in lymphocytes of persons living in West European regions contaminated by Chernobyl releases; dics = dicentric; cr = centric rings
a mean elevation

Region	Sample	Date of study	Method	[a]Results	Reference	Remarks
Salzburg Austria	17 adults	1987	dics+cr	Ca. 4-fold	Pohl-Rüling et al. 1991	
Germany southern regions	29 children + adults	1987-1991	dics+cr	Ca. 2.6-fold	Stephan, Oestreiche r 1993	physical dose estimate <0.5mSv
Norway: selected regions	44 Reindee r Samis, 12 sheep farmers	1991	dics+cr	10-fold	Brogger et al. 1996	physical dose estimate 5.5 mSv

Conclusions

The data in tables 2-6 show that neither the exposure of populations affected by the Chernobyl accident nor the effects have been evaluated correctly up to now. It must be discussed, however, that some of the authors interpret their somatic findings as genetically induced. In cases of continual exposure by radioactivity it is, indeed, not always certain whether the damage occurred *in utero* or by pre-conceptional mutation. Intrauterine deaths and infant deaths, as well as malformations, are also inducible by irradiation of the germ-cells in fathers or mothers. Rugh (1962) has already stated that the appearance of certain malformations of the brain are quite similar after pre-conceptional and *in utero* irradiation.

The studies of the A-bomb survivors did not show significant genetic effects. Investigations in the offspring of a large cohort of Sellafield workers, however, which were initiated after the leukaemia findings in the proximity of this British nuclear reprocessing plant, confirm a rise in stillbirths and congenital anomalies caused by pre-conceptional exposure of the fathers (Parker et al. 1999). The cited Chernobyl studies may therefore include some overlap of trans-generational and teratogenic effects. Although this conclusion demands further scientific effort for differentiation, the foetus must be considered as vulnerable to low dose exposures as was assumed in the early times of radiation research.

References

Akar, N., Cavdar, A.O., and Arcasoy, A., 1988, High incidence of Neural Tube defects in Bursa, Turkey, *Paediatric and Perinatal Epidemiol.* **2**:89-92.

Akar, N., Ata, Y., and Aytekin, A.F., 1989, Neural Tube defects and Chernobyl? *Paediatric and Perinatal Epidemiol.* **3**:102-103.

Bochkov, N.P., and Katosova, L.D., 1994, Analysis of multiaberrant cells in lymphocytes of persons living in different ecological regions, *Mutat. Res.* **323**:7-10.

Bogdanovich, I.P., 1997, Comparative analysis of the death rate of children, aged 0-5, in 1994 in radiocontaminated and conventionally clean areas of Belarus, in: *Medicobiological effects and the ways of overcoming the Chernobyl accident consequence,* Collected book of scientific papers dedicated to the 10th anniversary of the Chernobyl accident, Minsk-Vitebsk, p. 4.

Brogger, A., Reitan, J.B., Strand, P., and Amundsen, I., 1996, Chromosome analysis of peripheral lymphocytes from persons exposed to radioactive fallout in Norway, *Mutat. Res.* **361**:73-79.

Caglayan, S. Kayhan, B., Mentesoglu, S., and Aksit, S., 1990, Changing incidence of neural tube defects in Aegean Turkey, *Paediatric and Perinatal Epidemiol.* **4**:264-268.

Czeisel, A.E.; and Billege, B., 1988, Teratological evaluation of Hungarian pregnancy outcomes after the accident in the nuclear power station of Chernobyl, *Orvosi Hetilap* **129**:457-462 (Hungarian).

Dolk, H., Nichols, R., and a EUROCAT Working Group, 1999, Evaluation of the impact of Chernobyl on the prevalence of congenital animalies in 17 regions of Europe, *Int. J. Epidemiol.* **28**:941-948.

Ericson, A., and Kallen, B., 1994, Pregnancy outcome in Sweden after Chernobyl, *Environ. Res.* **67**:149-159.

Fritz-Niggli, H., 1997, *Strahlengefährdung/Strahlenschutz,* 4th ed., Hans Huber, Bern, Switzerland.

Godlevsky, I., and Nasvit, O., 1998, Dynamics of health status of residents in the Lugnyny district after the accident of the ChNPS, in: *Research activities about the radiological consequences of the Chernobyl NPS accident and social activities to assist the sufferers by the accident,* T. Imanaka, ed., Research Reactor Institute, Kyoto University, KURRI-KR-21, pp.149-156.

Grosche, B., Irl, C., Schoetzau, A., and van Santen, E., 1997, Perinatal mortality in Bavaria, Germany, after the Chernobyl reactor accident, *Rad. Environ. Biophys.* **36**:129-136.

Güvenc, H., Uslu, M.A., Güvenc, M., Ozkici, U., Kocabay, K., and Bektas, S., 1993, Changing trend of neural tube defects in Eastern Turkey, *J. Epidemiol. Community Health* **47**:40-41.

Harjulehto, T., Aro, T., Rita, H., Rytomaa, T., and Saxen, L., 1989, The accident at Chernobyl and outcome of pregnancy in Finland, *Brit. Med. J.* **298**:995-997.

Harjulehto, T., Rahola, T., Suomela, M., Arvela, H., and Saxén, L., 1991, Pregnancy outcome in Finland after the Chernobyl accident. *Biomed & Pharmacother* **45**:263-266.

Hille, R., Wolf, U., Fender, H., Arndt, D., and Antipkin, J., 1995, Cytogenetic examination of children in Ukraine, In *Radiation Research 1895-1995. Proceed. 10th Int. Congr. Radiation Res. Würzburg,* U. Hagen et al., ed., Germany.

Hoffmann, W., and Schmitz-Feuerhake, I., 1999, How radiation-specific is the dicentric assay? *J. Exp. Analysis Environ. Epidemiol.* **2**:113-133.

Jacobson, L., 1966, Radiation-induced effects in mouse foetuses, *Acta radiol.* **254**:82

Kolominsky, Y., Igumnov, S., and Drozdovitch, V., 1999, The psychological development of children from Belarus exposed in the prenatal period to radiation from the Chernobyl atomic power plant, *J. Child. Psychol. Psychiatry* **40**:299-305.

Kondrashenko, V.G. et al., 1996, Mental disorders caused by Chernobyl, in: *Report on the 3rd Int. Congress "World after Chernoby",* Minsk, 1996, cited in Nesterenko, V.B., *Chernobyl Accident. Radiation Protection of Populatio,* Republic of Belarus Institute of Radiation Safety "Belrad", Minsk 1998.

Korblein, A., and Kuchenhoff, H., 1997, Perinatal mortality in Germany following the Chernobyl accident, *Rad. Environ. Biophys.* **36**:3-7.

Körblein, A., Säuglingssterblichkeit nach Tschernobyl, 2003a, *Berichte Otto Hug Strahleninstitut* **24**:6-34.

Körblein, A., 2003b, Strontium fallout from Chernobyl and perinatal mortality in Ukraine and Belarus, *Radiats. Biol. Radioecol.* **43**:197-202.

Körblein, A., 2004, Fehlbildungen in Bayern nach Tschernobyl, *Strahlentelex No.* **416-417**:4-6.

Kordysh, E.A., Goldsmith, J.R., Quastel, M.R., Poljak, S., Merkin, L., Cohen, R., and Gorodischer, R., 1995, Health effects in a casual sample of immigrants to Israel from areas contaminated by the Chernobyl explosion, *Environ. Health Persp.* **103**:936-941.

Kozlova, I.A., Niagu, A.I., and Korelev, V.D., 1999, The influence of radiation to the child mental development, *Zh. Nevrol. Psikhiatr. Im. SS Korsakova* **99**:12-16 (Russ.).

Kruslin, B., Jukic, S., Kos, M., Simic, G., and Cviko, A., 1998, Congenital anomalies of the central nervous system at autopsy in Croatia in the period before and after Chernobyl, *Acta Med. Croatica* **52**:103-107.

Kulakov, V.I., Sokur, T.N., Volobuev, A.I., Tzibulskaya, I.S., Malisheva,V.A., Zikin, B.I., Ezova, L.C., Belyaeva, L.A., Bonartzev, P.D., Speranskaya, N.V., Tchesnokova, J.M., Matveeva, N.K., Kaliznuk, E.S., Miturova, L.B., and Orlova, N.S., 1993, Female reproduction function in areas affected by radiation after the Chernobyl power station accident, *Environ Health Persp.* **101, Suppl. 2**:117-123.

Lazjuk, G.I., Nikolaev, D.L., and Novikova, I.V., 1997, Changes in registered congenital anomalies in the Republic of Belarus after the Chernobyl accident, *Stem Cells* **15, Suppl. 2**:255-260.

Lotz, B., Haerting, J., and Schulze, E., 1996, Veränderungen im fetalen und kindlichen Sektionsgut im Raum Jena nach dem Reaktorunfall von Tschernobyl, Oral presentation at the International Conference of the Society for Medical Documentation, Statistics, and Epidemiology, Bonn, Germany (available on request).

Lüning, G., Scheer, J., Schmidt, M., and Ziggel, H., 1989, Early infant mortality in West Germany before and after Chernobyl, *Lancet II*:1081-1083.

Michel, C., Fritz-Niggli, H., 1977, Radiation damage in mouse embryos exposed to 1 rad X-rays or negative pions, *Fortschritte Röntgenstrahlen* **127**:276-280.

Mocan, H., Bozkaya, H., Mocan, Z.M., Furtun, E.M., 1990, Changing incidence of anencephaly in the eastern Black Sea region of Turkey and Chernobyl, *Paediatric and Perinatal Epidemiol.* **4**:264-268.

Moumdjiev, N., Nedkova, V., Christova, V., Kostova, Sv., 1992, Influence of the Chernobyl reactor accident on the child health in the region of Pleven, Bulgaria, in: *20th Int. Congr. Pediatrics Sept. 6-10, 1992 in Brasil*, p.57. Cited by Akar, N., Further notes on neural tube defects and Chernobyl (Letter), *Paediatric and Perinatal Epidemiol.* 8, 1994, 456-457.

Ohzu, E., 1965, Effects of low-dose X-irradiation on early mouse embryos, *Radiat. Res.* **26**:107.

Parker, L., Pearce, M.S., Dickinson, H.O., Atkin, M., Craft, A.W., 1999, Stillbirths among offspring of male radiation workers at Sellafield nuclear processing plant, *Lancet* **354**: 1407-1414

Petrova, A., Gnedko, T., Maistrova, I., Zafranskaya, M., Dainiak, N., 1997, Morbidity in a large cohort study of children born to mothers exposed to radiation from Chernobyl. *Stem Cells* **16**, Suppl. 2: 141-150

Pohl-Rüling, J., Haas, O., Brogger, A., Obe, G., Lettner, H., Daschil, F., Atzmüller, C., Lloyd, D., Kubiak, R., and Natarajan, A.T., 1991, The effect on lymphocyte chromosomes of additional burden due to fallout in Salzburg (Austria) from the Chernobyl accident, *Mutat. Res.* **262**:209-217.

Ponomarenko, V.M., Nagornaia, A.M., Proklina, T.L., Litvinova, L.A., Stepanenko, A.V., Sytenko, E.R., and Osnach, A.V., 1993, The morbidity in preschool children living on the territory of Rovno Province subjected to radioactive contamination as a result of the accident at the Chernobyl Atomic Electric Power Station, *Lik Sprava*. **2-3**:36-38 (Russ.).

Ramsay, C.N., Ellis, P.M., and Zealley, H., 1991, Down's syndrome in the Lothian region of Scotland – 1978 to 1989, *Biomed. Pharmacother.* **45**:267-272.

Roux, C., Horvath, C., and Dupuis, R., 1983, Effects of pre-implantation low-dose radiation on rat embryos, *Health Phys.* **45**:993-994.

Rugh, R., and Grupp, E., 1959, X-irradiation exencephaly, *Am. J. Roentgenol. Radium Ther.Nucl. Med.* **81**:1026-1052.

Rugh, R., 1962, Major radiobiological concepts and effects of ionizing radiations on the embryo and fetus, in Haley, T.J. and Snider, R.S., ed., Response of the nervous system to ionizing radiation, Academic press, New York, London, p. 3-26

Salomaa, S., Sevan`kaev, A.V., Zhloba, A.A., Kumpusalo, E., Mäkinen, S., Lindholm, C., Kumpusalo, L., Kolmakow, S., and Nissinen, N., 1997, Unstable and stable chromosomal aberrations in lymphocytes of people exposed to Chernobyl fallout in Bryansk, Russia, *Int. J. Radiat. Biol.* **71**:51-59.

Savchenko, V.K., 1995, The Ecology of the Chernobyl Catastrophe. Scientific outlines of an international programme of colloborative research. *Man and the Biosphere Series Vol. 17*, UNESCO Paris, p.83.

Scheid, W., Weber, J., Petrenko, S., and Traut, H., 1993, Chromosome aberrations in human lymphocytes apparently induced by Chernobyl fallout, *Health Phys.* **64**:531-534.

Scherb, H., and Weigelt, E., 1999a, Spatial-temporal logistic regression of the cesium contamination and the time trends in annual stillbirth proportions on a district level in Bavaria, 1980-1993, in: *Proceedings of the 14th international workshop an statistical modelling,* H. Friedl et al., ed., Technical University Graz, Graz, Austria, pp.647-650.

Scherb, H., Weigelt, E., and Brüske-Hohlfeld, I., 1999b, European stillbirth proportions before and after the Chernobyl accident, *Int. J. Epidemiol.* **28**:932-940.

Scherb, H., Weigelt, E., Brüske-Hohlfeld, I., 2000a, Regression analysis of time trends in perinatal mortality in Germany, *Environ. Health Persp.* **108**:159-165.

Scherb, H., Weigelt, E., 2000b, Spatial-temporal change-point regression models for European stillbirth data, in: *30th Ann. Meeting Europ. Soc. Radiat. Biol.,* Warszawa, Poland, August 27-31, Abstracts, p.14. http://www.tschernobylhilfe.ffb.org/dokumente/warschau.pdf

Scherb, H., and Weigelt, E., 2003, Congenital malformation and stillbirth in Germany and Europe before and after the Chernobyl nuclear power plant accident, *Environ. Sci.& Pollut.Res.* **10 Special** (1):117-125.

Scherb, H., and Weigelt, E., 2004, Cleft lip and cleft palate birth rate in Bavaria before and after the Chernobyl nuclear power plant accident, *Mund Kiefer Gesichtschir.* **8**:106-110 (in German).

Sevan´kaev, A.V., Tsyb, A.F., Lloyd, D.C., Zhloba, A.A., Moiseenko, V.V., Skrjabin, A.M., and Climov, V.M., 1993, ´Rogue´cells observed in children exposed to radiation from the Chernobyl accident, *Int. J. Radiat. Biol.* **63**:361-367.

Shidlovskii, P.R., 1992, General morbidity of the population in districts of the Brest region. *Zdravoohranenie Belorussii (Minsk)* **1**:8-11 (Russ.).

Sperling, K., Pelz, J., Wegner, R.-D., Schulzke, I., and Struck, E., 1991, Frequency of trisomy 21 in Germany before and after the Chernobyl accident, *Biomed. Pharmacother.* **45**:255-262.

Sperling, K., Pelz, J., Wegner, R.-D., Dörries, A., Grüters, A., and Mikkelsen, M., 1994, Significant increase in trisomy 21 in Berlin nine months after the Chernobyl reactor accident: temporal correlation or causal relation relation? *Brit. Med. J.* **309**:158-162

Stephan, G., and Oestreicher, U., 1989, An increased frequency of structural chromosome aberrations in persons present in the vicinity of Chernobyl during and after the reactor accident. Is this effect caused by radiation exposure? *Mutat. Res.* **223**:7-12.

Stephan, G., and Oestreicher, U., 1993, Chromosome investigation of individuals living in areas of Southern Germany contaminated by fallout from the Chernobyl reactor accident, *Mutat. Res.* **319**:189-196.

Strahlentelex 55, 1989, Säuglinge starben vermehrt oder wurden tot geboren, Berlin, Germany, p. 6.

Sychik, S.I., and Stozharov, A.N., 1999, Analysis of morbidity of children irradiated in utero as a result of Chernobyl accident, *Zdravoohranenie Belorussii (Minsk)* **6**:20-22 (Russ.).

Sychik, S.I., and Stozharov, A.N., 1999, Estimation of the influence of prenatal irradiation on functional state of critical organs and systems in children at distant terms following the Chernobyl accident. *Radiats. Biol. Radioecol.* **39**:500-504 (Russ.).

Ulstein, M., Jensen, T.S., Irgens, L.M., Lie, R.T., and Sivertsen, E., 1990, Outcome of pregnancy in one Norwegian county 3 years prior to and 3 years subsequent to the Chernobyl accident, *Acta Obstet. Gynecol. Scand.* **6**:277-280.

UNSCEAR: United Nations Scientific Committee on the Effects of Atomic Radiation, 1986, Genetic and Somatic Effects of Ionizing Radiation, Report to the General Assembly, New York 1986.

UNSCEAR: United Nations Scientific Committee on the Effects of Atomic Radiation, 1988, Sources, Effcts and Risks of Ionizing Radiation, Report to the General Assembly, New York 1988.

Verschaeve, L., Domracheva, E.V., Kuznetsov, S.A., and Nechai, V.V., 1993, Chromosome aberrations in inhabitants of Byelorussia: consequence of the Chernobyl accident, *Mutat. Res.* **287**:253-259.

Zieglowski, V., and Hemprich, A., 1999, Facial cleft birth rate in former East Germany before and after the reactor accident in Chernobyl, *Mund Kiefer Gesichtschir.* **3**:195-199 (in German).

CHAPTER 7
Reflections of the Chernobyl Catastrophe on the Plant World:
Special and General Biological Aspects
D.M.Grodzinsky
Ukrainian National Academy of Sciences

Over the past two decades we have become increasingly aware of the long-term impact of environmental radioactivity resulting from the accident at the Chernobyl Nuclear Power Plant. Interest in the reaction of plants to the influence of chronic irradiation in low doses has quickened with time because of concerns relating to a possible degradation of biota in the areas contaminated with radionuclides.

In addition, a view of the radiobiological processes induced in plants by chronic irradiation should elucidate the main tendencies in the formation of late effects of irradiation. As this takes place we bear in mind that these late effects in plants could not be related to 'radio-phobia', as it is called, as there is a tendency to assign the cause of injuries observed after the Chernobyl catastrophe merely to a *fear* of irradiation. We have seen, since the accident, clear and diverse effects of irradiation in plants over time (1; 2).

Emissions containing a huge total amount of radionuclides from the destroyed reactor unit caused a vast radioactive anomaly over a large area. The doses of irradiation varied over a wide range where radionuclides, resulting from the Chernobyl catastrophe, were deposited. The surface activity of radionuclides reached hundreds of thousands of Curies per square km in the immediate vicinity of the blasted reactor unit (1).

The formation of dose loads in plants is, by nature, different at various time periods after the emergence of radionuclides within the ecosystems. We recognized various periods in dose accumulation by plants. The first period lasted from 26 April 1986 till 15th of May 1986 and differs from the others in that the main sources of irradiation were related to short-lived radionuclides contained in radioactive clouds. Radionuclides were deposited on the surfaces of leaves, soil and water. As this took place, layers of absorbed radionuclides were fused (*arrized*) onto the surfaces of plant organs. Most of this radioactivity was concentrated in "hot particles", as they are called, although there were radionuclides in different physicochemical forms, including ionic forms. The components of plant irradiation, during the first period, were external irradiation from radionuclides distributed on ambient objects as well as irradiation from radioactive materials adsorbed by the outer cell layers of plant organs. The external irradiation was largely connected with gamma rays, but the applicative irradiation consisted of various types of radiation - gamma rays, beta- and alpha particles. Some activity of soluble radionuclides was concentrated in the cells and tissue of plants as the cells metabolized them (3).

The accumulated local dose under hot particles was so huge that the cells died and some microfunnels became fused (*arrized*). Because of this, the hot particles held securely onto the surfaces of leaves and other parts of the plants. In addition to this, the hot particles held owing to gluey substances and trichomes on leaf and bud surfaces. Assessment of doses showed that the absorbed dose for critical organs, namely apical meristems in cereal plants growing in plots contaminated with radionuclides, ranged from 1.33 to 12 Gy in 1986 when the surface density of radionuclide contamination varied from 1,18 to 8,4 x 10[8] Bq per m[2] (2).

The second period of irradiation (after May 1986 - over the course of the growing season) has been related to radiation from long-lived and a few specific short-lived radionuclides. The main components of irradiation were due to radio nuclides absorbed onto the surfaces of plant organs and internal irradiation from radionuclides accumulated within cells and tissues. Finally, the third period is characterized by progressive increase in the fraction of internal radiation resulting from long-lived radionuclides accumulated by plants from soil through mechanisms of mineral nutrition (4).

On the whole, the dose load characteristics of different species of plants fluctuated. General sources of the components of the dose are shown below.

Vital forms of plant	*Sources of irradiation*
Perennial plants which were formed before 1986 and had developed leaves at the time of the accident	Acute irradiation during the first days after the accident. External irradiation from radioactive clouds and deposited radionuclides. Applicative irradiation *Critical organs:* Cells of young leaves, meristems in developing buds
Perennial plants which were formed before 1986 and had not developed leaves by the time of the accident.	Acute irradiation during the first days after the accident Applicative irradiation. External irradiation *Critical organs*: Meristems of buds
Annual plants which began development before the accident.	chronic external irradiation Applicative irradiation Chronic internal irradiation
Plants, which began their development after 1986	Chronic internal irradiation related to root intake of long-lived radionuclides Chronic external irradiation

The territory around the Chernobyl Nuclear Power Station, especially the 10-km zone is characterized by very high levels of dose. Conifers growing in some places of this territory died out shortly after the accident. This dead forest was named the "Red Forest". The lethal dose for conifers is approximated 80 -100 Gy (5).

However, lethal outcomes in plants were observed only in some areas contaminated by large amounts of radionuclide activity. Cumulative impacts of chronic irradiation in low doses are much more important for the comprehension, assessment and prognosis of the late effects of irradiation on human beings, other living entities and biocenosis.

Various plant reactions to the influence of chronic irradiation can be observed in the neighborhood of the Chernobyl Nuclear Power Station. Among these are a high death rate of plant apical meristems, multiple distortion of metabolism, cytogenetic effects and various morphological abnormalities. The types and frequencies of these effects vary with irradiation doses.

It is well to bear in mind that there is a difference in the responses of plants to internal and external irradiation. The effectiveness of internal irradiation related to the radionuclides deposited within cells and tissue is much higher compared with the external irradiation of the same rays. Table 1 gives the data illustrating the impacts of internal and external irradiation on frequencies of gene reversion in barley pollen (6; 7).

Table 1   Frequencies of waxy reversions in barley pollen as a result of 55 days of exposure to various levels of radionuclides released at Chernobyl and a pure gamma field.

Dose rate, $\mu Sv.h^{-1}$	Total dose, MSv	Total reversion frequency per 10^6 pollen grains	Radiation-induced reversions per 10^6 pollen grains
Radionuclide contaminated conditions			
Control (0.96)	1.3	174	0
59	75	226	52
320	422	837	663
400	528	1235	1061
515	680	1705	1531
Chronic gamma irradiation			
Background (0.11)	0.1	82	0
5	3.0	145	63
50	29.6	150	68
500	296	198	116
5000	2960	192	110
50000	29600	292	210

The large distinction of irradiation effects between internal and external irradiation can be attributable to at least two causes. On the one hand, the relative biological effectiveness of internal irradiation by virtue of peculiarities of microdosimetry in tissue and cells can be much higher than in the case of external irradiation and, on the other hand, low doses of irradiation can be perceived as a certain alarm signal. There is abundant evidence supporting the effect of irradiation dose as a signaling factor in the response of plants to low doses of gamma irradiation. This factor induces active responses in plants directed towards an adaptation to chronic irradiation. It appears that there are two adaptive strategies to stress impacts in plants, namely, ontogenetic and population or phylogenetic adaptation. The first type of adaptive strategy is revealed by radioadaptation and resides in an augmentation of radioresistance after irradiation in low doses. The second type of adaptive strategy lies in an increase in frequency of genetic diversification, which enlarges the possibilities for active natural selection.

The mechanism of ontogenetic adaptation involves an induction of synthesis of additive enzymes concerned with DNA reparation. We have suggested that ontogenetic radioadaptation is a nonspecific response of plants and the synthesis of additive reparative enzymes may result from the action of not only ionizing radiation, but also by factors of another nature, for instance UV-B (8). Most likely the ontogenetic adaptive strategy operates as a source for sustaining the organism within modified environmental conditions marked by an increase in genotoxicity. Widening the diversification of plant forms in future generations is achieved by means of genome instability in response to irradiation in low doses.

It is reasonably safe to suggest that many of the plant responses to low level chronic irradiation may be considered as a consequence of the active reactions of plants associated with the realization of adaptive strategies.

Teratogenicity in Plants in the Zone of low dose chronic Irradiation

A marked alteration of morphogenesis in plants growing within territories contaminated by radionuclides has been noted. Commonly encountered radiomorphoses derives from stochastically distributed cells, which have lost the capability for division. Inactivated cells restrict the normal behavior of intact cells in respect of division and elongation. Malformations are developed as a consequences of these restrictions. Examples of radiomorphoses in oak and horse chestnut leaves are shown in Fig.1 and 2.

Fig. 1 Giant leaf of irradiated oak (*Quercus robur*) with some deformation of the leaf plate (right) and normal leaf from unirradiated tree.

Fig. 2. Typical radiomorphose in horse-chestnut (*Aesculus hippocastanum*)

It can be seen clearly that the formation of leaves in these trees vary considerably in shape. However, there exist two types of abnormalities of morphogenesis. Altered organs are occasionally characterized by a regularity of structure – these differ in structure considerably to normal organs but are rigorously regular. The other type of altered organs are deformed in their structure without any regularity. The "regular radiomorphoses" were precisely the forms more frequently encountered in plants growing within contaminated territories. In fig. 3 gigantic needles in spruce (*Picea abies*) are shown. The metamorphosed needles are located in the upper part of the stem. The sizes of these needles were eight-ten times larger then in the unirradiated control plants.

Fig.3. Metamorphosed needles in spruce (*Picea abies*) grown in the Alienated Zone of the Chernobyl Nuclear Power Station. Twig with gigantic needles in the upper part (left), Normal and metamorphosed twigs (right).

Accretion of plant organs, notably flowers, were of frequent occurrence. Remarkably, this has occured in parts of the plant which do not normally branch out, for example: in the pedicle of the common dandelion (*Taraxacum officinale*).

It seems likely that the extent to which the morphological malformations can be manifested does not depend on dose rate, but the frequency of its formations are strongly dependant on the overall doses of chronic irradiation. We observed very expressive and cognate morphological malformations in plants growing in Kiev where the dose rates were substantially below those in the Alienated Zone (2).

In this respect the manifestations of morphological malformations are similar to the stochastic process when deviations of growth from the norm follow the all-or-none principle.

Fig.4. Pine (*Pinus silvestris*) with a great excess of apices.

This type of morphological anomaly owes its origin to the deterioration of the mechanism of apical dominance. Excess of ramification observed in many plant species is conditioned also by this reason. The effect of an excess of "unscheduled" tops is seen clearly in Fig.4. An excess of lateral buds in irradiated spruce are shown in figure 5.

Fig.5. Extraordinary localization of buds in spruce twig.

One observation, which may be a guide to the nature of this mechanism, is that the gravity perceiving organ appears to be impaired - resulting in the loss of normal orientation. (6).

Fig. 6. Loss of gravitational orientation in spruce (*Picea abies*) Surviving in the Chernobyl Zone.

The cross section of needles was also altered, becoming round rather than kidney-shaped. A number of needles in one fascicle were larger than the norm. On occasion, a great excess of needles were formed and the stem was cluttered with a dense brush of needles (Fig. 7)

Fig.7 Dense location of extraordinary dying needles in pine

"Brushes" of this kind were located in various parts of the stem, usually in the apical tops. This interesting phenomenon is the result of a proliferation of the primordia of the scales attempting to form full needles. It is not an example of "radiophobia".

Gigantism in leaves has been accompanied by augmentation in the sizes of anatomical elements. This is shown in Fig 8.

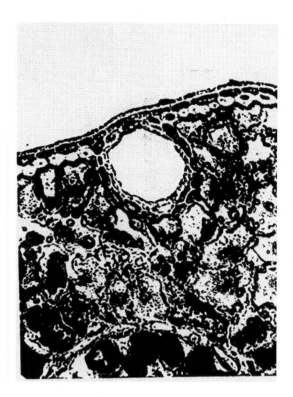

Fig. 8. Anatomical structure of abnormally enlarged needles in spruce.

We photographed regular radiomorphoses in various perennials, as well as in annual plant species in the first years after the accident at the Chernobyl Nuclear Power Plant, but morphological anomalies can also be observed in the Alienated Zone in recent years. Table 2 demonstrates the characteristics of needles in conifers in recent years (9).

Table 2. Morphometric indicators of control and chronically irradiated pine (*Pinus silvestris*) and spruce (*Picea abies*)

Plant	Median length of needles, mm	Median weight of needle, mg
Pine, control	60±4	80±3
Pine, chronic irradiation	19±3	14±2
Spruce, control	16±2	5±1
Spruce, chronic irradiation	40±3	95±5

Regular morphological abnormalities in plant structure may be classified in the following manner:

Type of morphological abnormalities	Primary causes of abnormalities
Irregular radiomorphoses	Proliferation of dead of meristematic cells stochastically distributed in tissue
Gigantism in leaves	Alteration of the positional control in meristem (increased number of quantal mitosis)
Dwarfism in leaves	Alterations of positional control in meristem (decreased number of quantal mitosis)
Polyapices	Removal of apical dominance in apical meristem
Formation of extraordinary apical and lateral buds	Slackening of apical dominance and alterations of positional control in the secondary meristem
Formation of additional extraordinary needles in the stems of conifers	Loss of positional control in cataphyls
Loss of geotropic orientation of organs	Slackening of polarity in organs
The surplus of ramification	Slackening of apical dominance

Alterations of physiological and biochemical processes were also observed in plants growing in circumstances where soil was contaminated with radionuclides. By way of illustration we refer to changes of colour in thyme. It is evident that the colour of a flower is dependant on the acidity of cell sap and from this follows the conclusion that the acidity of the cell sap changes under chronic irradiation. It is not inconceivable that this response of irradiated thyme also has a stochastic nature.

Mutagenic impact of irradiation in low doses on plants
Two general types of radiobiological effects in plants are recognized in cases of acute irradiation as well as in the event of low dose chronic irradiation: non-stochastic or deterministic and stochastic effects. Non-stochastic effects consist of somatic damages and functional deficiencies. Stochastic effects are characterized by the probabilistic nature of its manifestation. These effects have no dose threshold and the intensity of its manifestation does not depend on the dose. Mutagenesis of somatic and reproductive cells is referred to as a stochastic effect of irradiation (3).

The frequency of chlorophyll mutations was also increased by irradiation due to radionuclide contamination (Table 3)

Table 3. Yield of albina chlorophyll mutation (%) in rye and barley
seedlings in the 30 km zone

Plant	Control	Years of planting in contaminated soil			
		1986	1987	1988	1989
Rye (var. Kiev-80)	0.01	0.14	0.40	0.91	0.71
Rye (var. Kharkow-03)	0.02	0.80	0.99	1.20	1.14
Barley (var. No 2)	0.35	0.81	0.63	0.70	0.71

Activity of soil: ^{134}Cs, ^{137}Cs, ^{144}Ce, ^{106}Ru *and others, 180 kBq.kg^{-1}*

The genetics effects of irradiation can be demonstrated by using some mutant
plants. In these there is an emergence of reversion forms (tables 4 – 6).

Table 4. Influence of gamma irradiation on the frequency of reversions in the mutant form
of mouse-ear cress (*Arabidopsis thaliana f. Glabrous*)

Dose of irradiation	Number of leaves with trichomes	Cladding of leaves with trichomes, %
Control, unirradiated plants	0	
Acute irradiation, 10 Gy	3.8+0.7	1.9 + 0.1
Chronic irradiation, 0.3 Gy	9.7 + 1.8	1.9 + 0.3
Chronic irradiation, 0.5 Gy	11.7 +1.9	2.9 + 1.5
Chronic irradiation, 5 Gy	12.5 + 1.5	19.6 + 3.1

Table 5. Damage to cells of apical root meristem in onion (*Allium cepa*) growing in soil
contaminated with radionuclides

Specific radioactivity of soil, kBq.kg^{-1}	Number of cells under study	Mitotic index, %	Number of cells with aberrations, % in reference to control	Number of cells with micronuclei % in reference to control	Number of degenerating cells % in reference to control
Control	15005	4.1	100	100	100
37	33275	4.4	240	171	250
185	29290	4.4	216	129	500
370	23325	117	150	229	900

Table 6. The manifestation frequency of various polyploid pollen cells in mouse-ear cress (*Arabidopsis thaliana*)) under chronic irradiation (the first vegetation), %

Ploidy of pollen grain	Control, unirradiated plants	Dose, Gy	
		0.5	5
Diploid cells	91	77	65
Tetraploid cells	1	9	15
Chimaeric tissue	8	14	20

Data presented in these tables point to the fact of a strong influence of chronic irradiation on cytogenetic distortions in plant cells. The effects of chronic irradiation embrace the various genetic events from genome to point mutations.

Alterations in metabolic processes in irradiated plants

Intensities of many biochemical processes are changed in plants surviving in the Chernobyl zone. The infringements of metabolic reactions are basically linked to a violation of signaling systems in plants as well as protection reactions to stress. The previously mentioned reactions are unspecific, being realized when stress factors of different a nature influence plants. This phenomenon is illustrated by data on an augmentation of anthocyanin concentration in plants irradiated with gamma rays or growing in a radionuclide contaminated area (table 7)

Table 7. Content of anthocyanins in irradiated plants

Plant species	Type of irradiation	Content of anthocyanins, % in reference to control
Maize (*Zea mays*), seedlings	Control	100
	Acute irradiation, 10 Gy	127
	Chronic irradiation: plants were grown in soil contaminated with radionuclides, 975 Bq.kg^{-1}	119
Bean (*Phaseolus aureus*), hypocotyl	Control	100
	Acute irradiation, 10 Gy	123
	Chronic irradiation 0.5 Gy	157
Mouse-ear cress (*Arabidopsis thaliana f.Columbia*)	Control	100
	Acute irradiation, 10 Gy	169
	Chronic irradiation, 0.5 Gy	173

As is obvious from these data the effect of chronic irradiation is greater than in the controls.

Tumor formation in plants grown under low dose irradiation

Inasmuch as processes of growth and morphogenetic regulation as well as positional control in plants are injured by irradiation, there is reason to assume that there are bound to be reductions in the normal response to the stimuli provided by the bacterial tumor

Agrobacterium tumefaciens. Stimulation of growth of these tumors in plants does occur under the action of low-level irradiation. This has been demonstrated in experiments on explants from potato tubers grown in soil extracts contaminated with radionuclides. The specific radioactivity of these extracts was approximately 3 kBq.l^{-1}. Stimulation of growth in callus culture from *Datura stramonium* has been also established (3). Table 8 shows tabulated data illustrating the impact of chronic irradiation on tumor tissue in jimson-weed.

Table 8. Impact of soil extracts on growth and cell divisions in normal and tumor tissue of *Datura stramonium*.

Content of variant	Weight of callus,		The average Number of cells in 1 g of tissue, x 10^5		The average number of cells in callus, x 10^5	
	G	%		%		%
Control, normal tissue	1.98	100	39.7	100	78.6	100
Normal tissue treated with extract from radioactive soil	2.58	130.2	38.9	98	100.4	127.6
Tumor tissue	.24	100	23.0	100	74.5	100
Tumor tissue treated with extract from radioactive soil	2.82	87.2	32.4	140.7	91.5	122.8

(Radioactivity of soil –3.1 10^4 Bq.kg$^{-1,}$ radioactivity of soil extract – 20 Bq.l^{-1}. Soil was contaminated with ^{137}Cs and ^{144}Ce)

New tumor-like formations arise in many plant species *(Hieracium murorum, Hieracium umbellatum, Rubus idaeus, Rubus caesius* and others) grown in natural conditions on territories contaminated with radionuclides. Tumor-like nodes were found in nearly 80 %. Of *Sonchus arvense* grown in contaminated soil (3).

An increase in gall formations was observed in leaves of oaks in forests contaminated with radionuclides.

Induction of genome instability in plants

There is evidence for active cellular processes initiated by chronic irradiation in low doses in plants. The display of the range of phenomena, namely genomic instability, bystander effects and adaptive responses has convinced radiobiologists to consider a new supposition regarding the contribution of radiobiological responses from active cellular processes.

Genome instability is a term used to describe an increased rate of genetic alterations. Genomic instability, with respect to chromosomal endpoints, was first described as the delayed onset of *de novo* chromosomal aberrations (10) and induction of mutations.

We have observed radiation-induced chromosomal instability in many species of plants after their transference from radioactive soil to "Clean" substrate. These studies have involved continuing exposure to a wide range of dose rates and observations on the late effects of irradiation (Table 9).

Table 9. Yield of chromosome aberrations (%) by chronic irradiation in root apical meristems of different plant species*

Plant species	Control	1986	1987	1988	1989
Lupinus alba	0.9	19.4	20.9	14.0	15.9
Pisum sativum	0.2	12.9	14.1	9.1	7.9
Secale cereale	0.7	14.9	18.7	17.1	17.4
Triticum aestivum	0.9	16.7	19.3	17.7	14.2
Hordeum vulgare	0.8	9.9	11.7	14.5	9.8

* Seedlings were grown in a nutritive solution containing 70,000 Bq.l^{-1}

Genome instability induced by chronic irradiation in low doses can be best observed in the behavior of winter wheat plants in our "Collection of Chernobyl Wheat Mutants". This collection originated with genetically modified plants gathered in 1987 in the neighborhood nearest to the wrecked reactor unit (11). Fields of winter wheat were located in territories characterized by high levels of radioactivity. The plants were exposed to severe irradiation from radioactive clouds, radionuclides were absorbed into the surfaces of organs in 1986 and from inner sources of radiation due to incorporated radionuclides in 1987. Ears of wheat have fallen and compact families originating from each ear developed in the autumn of 1986. There were many morphologically changed forms. The famous wheat selectionists Prof. Shkvarnikov P.K. and Prof. Batygin N.F. gathered seeds of severely modified wheat plants. These seeds were sown in clean territory in the autumn of 1987. Beginning in that year and up to this point, the genetic behavior of these plants has been exemplified by an uninterrupted, continuous change. Variability of the morphological characteristics is very wide, year after year, covering ear form, height of ears and plants, presence of awnes, glumes, physiological and biochemical peculiarities, including ripeness, quality of gluten and other characteristics. The variability of mutants in the collection is very intensive and unpredictable.

In Table 10 frequency of the main morphological anomalies in winter wheat are demonstrated.

Table 10. The frequency of morphological anomalies in three generations of winter wheat grown over two years in highly contaminated fields

Type of morphosis	Years of investigation		
	1986	1987	1988
Zones of sterility in ear	49.0	29.8	1.9
Shortening of ear	10.0	9.4	0.8
Ramification of ear	4.5	11.1	9.4
Shortening of stem	4.5	5.7	4.9
Lengthening of stem	4.4	4.7	5.4
Augmentation of awns	2.8	2.8	4.7
Unevenness of awns	1.4	3.4	2.9
Square-headed forms	4.9	14.0	24.7
Altered colour of stem	0.9	1.7	1.9
Gigantism in ear	1.4	1.8	2.9
Additional earlets	14.0	14.8	29.7

The extensively widespread morphological modifications among the plants in the collection are shown in Fig. 9.

Fig 9. Diversified forms of ears in winter wheat showed evidence of significant genome instability.

Radiation-induced genome instability is apparently connected with a specific mechanism. Several groups have demonstrated that proteins associated with the repair of DNA may contribute to the instability process, such as DNA-Pros and p53 (10). Certain DNA repair proteins have been shown to take part in the protection of telomeres (10). Cells having mutations in genetic coding for these proteins are subject to multiple forms of instability - defective double-strand break repair, chromosomal end-to-end fusion, and joining of unprotected telomeres to the radiation-exposed ends of the double-strand breaks. It is conceivable that epigenetic alterations (changes in methylation, acetylation, and phosphorylation) may also be responsible for the induction of instability in the phenotype (10).

From these numerous facts it is apparent that radiobiological responses are manifested with high intensity on various levels of plant systems organization. Relative biological effectiveness of chronic irradiation is very high especially for internal irradiation. The types of effects from chronic irradiation differ in significant ways from radiobiological responses of plants under acute irradiation. Late effects of irradiation play an increasingly important part in the augmentation of risk for biota on the territories contaminated with radionuclides. These essentially genetic effects can be very dangerous for plant individuals and populations. It is not inconceivable that, aside from the injuring influence of chronic irradiation, there is a signal impact within this factor on living beings and a system of responses has to be manifested as an integral adaptive reaction. Induction of genome instability, changes of reaction norms, widening of the spectrum of phenotypes, an increase in the frequency of mitotic cross-over and induction of unscheduled apoptosis in meristem are all components of this perverse adaptive reaction of organisms.

The radiobiological effects, which have been investigated in plants growing in territories contaminated with radioactive substances, do not, indeed, pertain only to plants: They are applicable for all living beings, as common principles form the basis of these radiobiological responses. The menace from the Chernobyl Catastrophe has not grown dim: It is with us and will be around for a long time.

References

1. Chernobyl Catastrophe/ Ed. By V.G. Baryakhtar. - Kiev: Editorial House of Annual Issue "Export of Ukraine", 1977. -576 p.
2. Chornobyl. The Exclusion Zone. Collected Papers. (Ed. V.G. Baryakhtar. -Kiev: Naukova Dumka, 2001. -548 P. (in Russian).
3. Grodzinsky, D.M., Kolomietz, O.D., Kutlachmedov, Yu.A. Anthropogenic Anomaly and Plants. -Kiev: Naukova dumka, 1991, 160 P. (in Russian).
4. Grodzinsky, D.M. Consequences of the Chernobyl Catastrophe as a Prototype of Nuclear Terrorism. - Defense and the Environment: Effective Scientific Communication (Eds. K.Mahutova et al. Kluwer Academic Publishers, 2004. - P.119 - 137.
5. Kozubov, G.M., Taskaev, A.I. Radiobiology Investigations of Conifers in Region of the Chernobyl Disaster (1986 - 2001). Moscow: PPC "Design. Information. Cartography", 2002. -272 P.
6. Bubryak, I., Vilensky, E., Naumenko, V., Grodzinsky, D. Influence of Combined Alpha, Beta and Gamma Radionuclide Contamination on the Frequency of Waxy-reversions in Barley Pollen.// Sci. Total Environ., 1992, **112**, 29 - 36 P.
7. Grodzinsky, D.M. Late Effects of Chronic Irradiation in Plants after the Accident at the Chernobyl Nuclear Power Station// Radiat. Protect. Dosimetry, 1995, **62**, p. 41 - 43.
8. Danil'chenko O.O., Grodzinsky D.M. Participation of DNA reparation systems in forming of radiation adaptive reaction in plants / in: Paradigms of contemporary radiobiology; Radiation protection of personnel of the objects of atomic energetics. – Kiev: Chernobyl, 2004. P. 15 – 16.
9. Sorochinsky B.V. Peculiarities of protein contents in abnormal needles in spruce (Picea abies) and pine (Pinus silvestris) from 1-km Zone of Chernobyl Nuclear Power Plant// Cytology and Genetics, 1998, vol. 32, No 5, P. 35 – 40.
10. Kadhim, M.A., Moore, S.R., Goodwin, E.H. Interrelationships amongst Radiation-Induced Genomic Instability, Bystander Effects, and the Adaptive Response // Mutation Research, 2004, **568**, p. 21 - 32.
11. Grodzinsky D.M./ Kolomietz O.D., Burdenjuk L.A. Collection of Chernobyl Mutants of Winter Wheat. –Kiev< 1999. – 30 p. (In Ukrainian).

CHAPTER 8
Infant Leukaemia in Europe after Chernobyl and its Significance for Radioprotection; a Meta-Analysis of Three Countries Including New Data from the UK.

Chris Busby
University of Liverpool, UK
Green Audit, UK
christo@greenaudit.org; christo@liv.ac.uk

1. Background

The Chernobyl accident contaminated most of Europe (Savchenko 1995) with fission product radioisotopes including short-lived, high-activity Iodine and Tellurium, and also fuel particles containing uranium and other intermediate half-life isotopes, including the 30 year half life Caesium-137. In the UK, whole body monitoring showed the persistence of Caesium-137 in the population (Etherington and Dorrian 1991) and grassland surveys enabled the radiological modelling of equivalent dose. In general, the exposures in Europe were examined in some detail and doses to the population were well characterised. For all of the countries of Europe except Belarus, the first year average committed effective doses were well below 1 mSv and in the non-Soviet countries less than 0.3mSv. At these levels, the risk models of the International Commission on Radiological Protection (ICRP) predict no measurable health effects. The doses are less than a quarter the natural background dose, and if dose has any universal radiological meaning, the exposures must be considered safe. Nevertheless, there were reported health effects. The most clear and graphic set of effects was the reported increases in infant leukaemia in the *in utero* exposed cohort in Greece (Petridou et al, 1996) , Germany (Kaletsch et al 1997) and Wales and Scotland (Gibson et al 1988, Busby and Scott Cato 2000, 2001).

Busby and Scott Cato examined the likely doses to the children and applied the current radiation risk models of the ICRP, those employed also by all radiological protection legislation, to show that the risk factors currently being employed for the protection of members of the public were in error by upwards of 100-fold. Such an error would begin to explain other apparently inexplicable associations between childhood leukemia and exposure near nuclear sites, notably the ongoing child leukaemia cluster near the UK Sellafield reprocessing plant in Cumbria. The importance of the infant leukaemia results are that the doses were well characterised and that, since the cohort is so well described, there is really no other explanation for the finding. Thus the existence of the effect may be taken as *prima facie* evidence of the failure of the ICRP model and may be used to determine the accurate risk factors for this kind of internal exposure.

The seriousness of this matter led, in the UK, to the formation of the Committee Examining Radiation Risk from Internal Emitters CERRIE. This was set up in 2001 at the joint request of the UK Minister of the Environment (Mr Michael Meacher) and Minister of Health (Ms Yvette Cooper). CERRIE's remit was to examine the assertion that, for internal exposures from fission–product radioisotopes, the true risk factors for cancer and leukaemia were much greater than those currently employed by the radiation protection legislation. The ICRP model was largely based on the historical external radiation exposure studies, principally that of the Japanese A-bomb survivors. On the other hand, by 2003, the new European Committee on Radiation Risk (ECRR) had published its model, which attempted to deal pragmatically with the clear evidence from various epidemiological and theoretical sources that internal exposures could, in certain cases, represent enhanced risks of up to three orders of magnitude or more (ECRR2003).

As part of its remit to examine the issue, CERRIE applied to the Oxford-based Childhood Cancer Research Group (CCRG) in order to follow up the Busby and Scott Cato analysis (which was for Scotland and Wales) by examining the UK by contamination area and period. Data limitations had forced Busby and Scott Cato (2000) to employ very slightly different periods to those used by Petridou *et al.* (1996) and Kaletsch *et al* (1997) and CERRIE decided to obtain data for the same periods. The first question was whether there was an effect in the high and intermediate exposure areas of the UK if the time periods used by Petridou *et al.* were used to define exposure cohorts. Exposure in the UK depended upon rainfall at the time and areas were agreed on the basis of measurements made by the UK National Radiological Protection Board and supplied to CERRIE. In this paper, because CCRG supplied a set of data that was for exactly the same cohort as that examined by Petridou *et al* in Greece and Kaletsch *et al* in Germany, I combine the three populations and weighted doses into one meta-analysis, which I employ to examine the risks of infant leukaemia from this type of internal exposure compared with the best available external exposure data, that of the Oxford Series obstetric X-ray studies (Wakeford and Little 2003).

The CCRG, Greek and German Data

To try and answer the questions relating to the UK exposures, numbers of infant leukemias (0-1yr) were obtained from the Childhood Cancer Research Group (CCRG) in Oxford for three areas and various periods before and after Chernobyl, together with total births occurring in these areas and periods. Two analyses were agreed. Both were aimed at examining the hypothesis that infant leukaemia rates had been increased following exposure *in utero* to radiation from the Chernobyl accident fallout in the UK. To this end, the UK was divided into three groups of areas considered to be high, intermediate and low exposure. Two sets of periods were also defined. The first was to enable an analysis of the periods used by Petridiou *et al.* in her analysis of the effects in Greece (Petridou *et al.*, 1996). The second was specified by CERRIE to examine a longer time period than that available to Petridou *et al.*. Although this second period analysis gave a larger effect, in this paper I wish only to examine the Petridou period, since I can combine these data with the earlier results.

In order to carry out such a meta-analysis I use data from Petridou *et al* and Kalestch *et al,* given in Table 4.

Table 1. Time periods for which data on infant leukaemia were supplied by CCRG

Cohort Group code	Time period	*In utero* exposure
Petridou *et al* analysis periods		
A	01/01/80 to 31/12/85	Unexposed
B	01/07/86 to 31/12/87	Exposed
C	01/01/88 to 31/12/90	Unexposed

Table 2. Data from CCRG on number of infant leukaemia cases.
1st analysis (time periods as Petridou) Males+Females

Period	Exposure category			
	High	Medium	Low	Total
01/01/1980 to 31/12/1985	3	52	66	121
01/07/1986 to 31/12/1987	1	16	24	41
00/01/1988 to 31/12/1990	2	39	35	76
Total	6	107	125	238

Table 3 Births in the analysis periods from CCRG
(time periods as Petridou) Males+Females

Period	Exposure category			
	High	Medium	Low	Total
01/01/1980 to 31/12/1985	90027	1783873	2363521	4237421
01/07/1986 to 31/12/1987	23152	479996	608921	1112069
00/01/1988 to 31/12/1990	47971	997941	1236102	2282014
Total	161150	3261810	4208544	7631504

Table 4 Infant leukaemia in UK Greece and Germany in the Chernobyl *in utero* exposure periods, (with rates per 100,000 and mean population-weighted foetal doses).

	Mean Dose [d](mSv)	Period A unexposed	Period B Exposed	Period C unexposed
[a]UK all cases	0.02	121 (2.86)	41 (3.69)	76 (3.33)
UK births		4237421	1112069	2282014
[b]Germany all cases	0.1	83 (2.30)	35 (3.76)	60 (2.96)
Germany births		3601176	928649	2029613
[c]Greece all cases	0.2	22 (2.75)	12 (7.35)	9 (2.89)
Greece births		801175	163337	311391
All 3 all cases	0.067	226 (2.62)	88 (3.99)	145 (3.13)
All 3 births		8639772	2204055	4623018

[a] *from CCRG;* [b]*from Kaletsch et al;* [c] *from Petridou et al.;* [d] *from original data, furnished by NRPB for CERRIE*

3. Method

Throughout this analysis I employ the same method that is used by the earlier researchers. I use standard contingency table analysis to compare unexposed (periods A + C) with exposed (B) cohorts. I combine all the UK categories; high, medium and low. I calculate relative risks, confidence intervals and significance based on Mantel Haenszel Chi-Square statistics for various groups in the UK and then for the combined dataset. I then examine the predictions of the risk factors derived from the obstetric X-ray series (Wakeford and Little 2003) with the excess risk found to calculate an error factor for the application of the obstetric X-ray risks and also the ICRP model risk.

4. Results

In the United Kingdom, the fallout from Chernobyl was patchy and related to outbreaks of thundery rain that occurred in Scotland, Wales and Yorkshire. Consequently, the low exposure areas defined for the CCRG data, mainly southern Britain, received little exposure. The high exposure areas were quite small in population and therefore I have combined the high and intermediate exposures groups for an examination of the effect in the UK, although for the combined analysis I use the entire UK dataset. In Table 5 I show the relative risk for the combined high and intermediate exposure cohorts in the periods of exposure relative to the period of no exposure. In Table 6 I combine all the UK groups with the German and Greek groups to give a comparison of the exposed and unexposed groups in the combined database of the three countries.

Table 5. Statistics of infant leukaemia rates in the UK based upon high + intermediate exposure groups in Scotland, North Wales and Yorkshire. Comparison of exposed (B) and unexposed (A+C) periods after Petridou *et al.*; data from CCRG.

Data Period	Cases High + Intermediate (rates)	Population High + Intermediate
A	69 (2.8)	2453548
B	25 (4.0)	632073
C	37 (2.9)	12840973
Statistics. B vs (A + C)	Relative Risk 1.4 (95% C. I. 0.88< RR< 2.20) $\chi^2 = 2.26$; p = 0.132; two tailed	

Table 6. Statistics of infant leukaemia in the combined population of UK, Germany and Greece using all UK data from CCRG plus the other data from Petridou *et al* 1996 and Kaletsch *et al* 1997

Data Period	Cases High + Intermediate (rates)	Population High + Intermediate
A	226 (2.62)	8639772
B	88 (4.0)	2204055
C	145 (3.1)	4623018
Statistics. B vs (A + C)	Relative Risk 1.43 (95% C. I. 1.13< RR< 1.80) $\chi^2 = 9.1$; p = 0.0025; two tailed	

5. Discussion

Errors in employing external radiation risk factors

In the UK data, supplied by CCRG, and based upon the Petridou *et al* 1996 birth cohort criteria, there was an increase in infant leukaemia in the exposed cohort in both the high and intermediate group and in the total population. Unlike the increases in Scotland and Wales found by Busby and Scott Cato 2000, this increase fell short of statistical significance at the p = 0.05 level using a two tailed test. This differs from the finding of Busby and Scott Cato mainly because different areas were employed by CERRIE and a slightly different period was employed. Combining the UK increases with those in Greece and Germany, (where the doses were greater) gave a 43% increase in infant leukaemia in the combined cohort of 2.2 million births in children exposed to a mean population weighted dose of 0.067 mSv. This dose is calculated from data supplied by NRPB to CERRIE and also based on UN data from Savchenko 1991 and also data from Kaletsch *et al*.1997. The mean dose was obtained by population weighting the foetal doses determined for each country supplied by NRPB to the CERRIE committee for UK and obtained from the German study where the doses were measured by the German Radiological Protection personnel and from UN data for Greece. I should emphasize that the internal dose here is unknown. The dose calculations are based mainly upon external dose, mainly gamma shine from Caesium-137 deposition. However, it is just this (mainly) external dose that is employed in radiological modelling of health effects and so, for the purpose of what follows, this is the dose that is relevant.

In calculating the dissonance between the predictions of the ICRP models and the observed number of cases found in Scotland and Wales, Busby and Scott Cato 2000 used the ICRP foetal risk factor of 0.0125 per Sievert (employed by COMARE 1996 to examine the Sellafield child leukemias). However, in discussions within CERRIE, it was pointed out that the obstetric data of Stewart *et a*l (1956) was a firmer basis on which to base any analysis of the risks from internal exposure. Stewart *et al* found a 40% increase in childhood cancer aged 0-14 after an X-ray dose of 10mSv (Wakeford and Little 2003). Although this has been disputed by some as being causal this value has been translated by Wakeford and Little into a relative risk of 50 per Sievert. It is this value that both Richard Wakeford of British Nuclear Fuels Ltd and Colin Muirhead of NRPB used to analyse the infant leukemias for the CERRIE report (CERRIE 2004a).

If I take it that a 10mSv X-ray dose causes a 40% increase in childhood cancer, we can see from Table 6 that a mean dose of 0.067 mSv from Chernobyl fallout has caused an increase in infant leukemia of 43%. The corresponding error in the application of the obstetric external risk factor to the infant leukemias is thus 43/40 * 10/ 0.067 = 160. There were therefore 160 times more infant leukemias in this combined population than would be predicted by the use of the obstetric X-ray data. And this is only in children aged 0-1. This is a minimum value, as we have yet to see what other cancers or leukemias emerge in this group as they age between 1 and 14 years. I am in the process of obtaining data from Scotland and Wales to enable me to carry out such an analysis.

Dose response relationship.
Because the database is so large, we can safely conclude that there was an increase in infant leukaemia in the high and intermediate exposure areas in those children who were exposed *in utero* to the fallout from Chernobyl. We cannot say that the effect was not due, in part, to parental pre-conception irradiation, since our exposed groups were born up to the end of 1987.

In addition to the discovery of a clear effect, there was also a biological gradient in the rates over a certain range. Fig 1 shows the increases in infant leukaemia in the European countries with dose. In Germany, Kaletsch *et al* 1997 had data for three dose areas and found that the dose response was biphasic. This is also true for the data from the UK when it is subdivided into the high, intermediate and low dose areas. In both countries the highest effect was in the intermediate dose area. Infant leukaemia increases were also reported in Belarus (Ivanov *et al*,) and the effect there was quite modest although the doses were higher than in Greece. The data suggest that over the range 0-2mSv the overall dose response is biphasic.

Fig 1 Dose response for infant leukaemia in the countries examined by this study and CERRIE. (Data from CCRG and CERRIE 2004a.) Effect is fractional excess risk, and dose is in mSv.

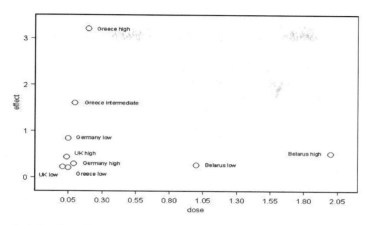

This biphasic behaviour is not so remarkable for an *in utero* cause and endpoints in the living child since above a certain dose some defence system may become overwhelmed and foetal death may intervene. However, it is uncertain that the dose levels reported correlate with internal exposures of the specific type that cause the illnesses. In the main, the exposures used for these studies are based upon external radiation measurements or ground deposition of Caesium. If the exposures were to milk from cattle fed winter silage contaminated with radionuclides, this milk might end up anywhere in the area, not necessarily where the main deposition was; indeed dairy cattle are unlikely to be feeding in such areas where the rainfall is high e.g. mountains.

The results found here show that there was an increase in infant leukaemia in the high and intermediate exposure categories chosen by CERRIE in the UK. The magnitude of the effect was lower than that found by Busby and Scott Cato whose diagnosis period study found RR = 3.87 p = 0.0001 for the combined Scotland and Wales exposed cohort, roughly in line with an earlier report by Gibson *et al.* for Scotland. This may not be surprising since the areas defined by the CERRIE analysis are quite different.

Given the extremely low mean doses involved in the combined exposure area, UK, Greece and Germany (< 70μSv), the increase in infant leukaemia was not predicted by the ICRP model and defines an error in the use even of the risk factor defined by the obstetric X-ray data of at least 160-fold.

The CERRIE committee analysis of the data found a similar mismatch between the expected number of cases in the UK, Germany and Greece but failed to draw any sensible conclusion. Table 4A6 from CERRIE 2004 is reproduced below as Table 7. The low value for Belarus may result from a biphasic dose response, but at least part of the explanation is the falsification of leukaemia records by the Soviet State who were concerned to minimise the evidence of the effects of the Chernobyl catastrophe (Busby and Yablokov 2006). Despite the clear dissonance between what was calculated by the representative of British Nuclear Fuels Richard Wakeford and the chief epidemiologist of NRPB, Colin Muirhead, who prepared table 4A6, the text of the CERRIE report (p88) is at odds with the table, stating clearly: *any suggestion that the risk of infant leukaemia arising from exposure to radionuclides in Chernobyl fallout has been materially underestimated is largely based on the findings of just one study.*

In the body of the majority CERRIE report the rapporteurs conclude:

Overal,l the findings . . . do not provide sufficiently persuasive evidence that the risk of internal exposure to radionuclides is seriously underestimated by risk estimates obtained by studies of exposure in utero to sources on internal irradiation. In the judgement of the large majority of the Committee members, it is likely that radioactive fallout from the Chernobyl accident resulted in an increased risk of infant leukaemia. A substantial fraction of members think that this increase is at the level anticipated by current risk models.

It is hard to see how these statements can be rationalised with the results presented in the table 4A6 in the same report. The International Cancer Agency in Lyon (IARC) had been expected to produce a report on infant leukaemia and wrote to CERRIE in 2004 to say that they were finishing off a study, but an enquiry at the time of writing resulted in no new news on this IARC investigation.

Table 7 Table 4A6 from CERRIE 2004a; Excess relative risk of infant leukaemia estimated in four birth cohorts.

Study	ERR (95% CI)	ERR coefficient Sv^{-1} (95% CI)	ICRP ERROR Ratio to external ERR coefficient (95% CI)
Great Britain	0.22 (-0.14, 0.69)	11000 (-7000, 34500)	220 (-140, 690)
Greece	1.6 (0.4, 4.1)	8000 (2000, 20500)	160 (40, 410)
Germany	0.48 (0.02, 1.15)	4800 (200, 11500)	96 (4, 230)
Belarus	0.26 (-0.24, 1.10)	130 (-120, 550)	2.6 (-2.4, 11)

Conclusion

The ICRP model has been criticised for lack of scientific method (ECRR2003). In particular, the ECRR argued that the use of acute external irradiation data to inform the model for health risks from internal chronic irradiation involved misuse of scientific method and employed deductive rather than inductive reasoning. More recently, the French nuclear risk agency IRSN has made largely the same points and has agreed that the ECRR criticisms are valid (IRSN 2006). If these concerns are valid, then clearly it is not possible to employ risk factors culled from the Japanese A-Bomb series to inform risk about internal irradiation. And, by the same argument, it is not valid to employ the risk factors obtained from the obstetric X-ray data. It is necessary to employ studies of children exposed to internal chronic radiation from fission product isotopes if we wish to develop models to predict or explain these same exposures. The nuclear site child leukemia clusters, e.g. Sellafield, Dounreay and La Hague, and others reviewed by ECRR2003 have been extensively studied and confirmed as being real and not due to chance. The denials of causality (e.g. COMARE 1996) have been based upon the external irradiation risk models. The difference between the yield of childhood leukaemia predicted by the ICRP or by the obstetric data and the observed numbers for these nuclear sites is in excess of 300-fold. The existence of infant leukaemia in these European cohorts shows that this denial is based upon an invalid model. It seems clear that this model is completely falsified by the existence of the infant leukemias in these European cohorts, clear and objective effects described by different groups in different countries. These increases in infant leukaemia have resulted from doses lower than those experienced by the nuclear site children who developed leukaemia. In the case of the Chernobyl infant leukemias there is no alternative explanation apart from internal radiation exposure to largely the same isotopes as the nuclear site leukaemias. The significance for this result for radiation protection is overwhelming.

References

Busby C and Scott Cato M (2001) 'Increases in leukemia in infants in Wales and Scotland following Chernobyl: Evidence for errors in statutory risk estimates and dose-response assumptions'. International Journal of Radiation Medicine 3 (1-2) 23

Gibson BE, Eden OB, Barratt A et al. (1988) 'Leukemia in young children in Scotland' Lancet 630

Busby C and Scott Cato M, (2000) 'Increases in leukemia in infants in Wales and Scotland following Chernobyl' Energy and Environment 11 (2) 127-137

CERRIE Minority Report (2004b) Minority Report of the Committee Examining Radiation Risk from Internal Emitters. (Aberystwyth: Sosiumi Press)

CERRIE Report (2004a) Report of the Committee Examining Radiation Risks from Internal Emitters.' (Chilton :NRPB)

COMARE, (1996) The Incidence of Cancer and Leukaemia in Young People in the Vicinity of the Sellafield Site in West Cumbria: Further Studies and Update since the Report of the Black Advisory Group in 1984, COMARE 4th Report (Wetherby: Department of Health).

ECRR (2003) 2003 Recommendations of the European Committee on Radiation Risk; Health Effects of Ionising Radiation Exposure at Low Doses for Radiation Protection Purposes Regulators' Edition: Brussels, 2003 Aberystwyth: Green Audit

G. Etherington and M. D. Dorrian (1991), `Radiocaesium levels, intakes, and consequent doses in a group of adults living in Southern England'. International Atomic Energy Agency Document IAEA-SM-306/29 (Vienna: IAEA).

IRSN (2005) Les consequences sanitaires des contaminations internes chroniques par les radionucleides. Avis sur le rapport CERI 'Etudes des effets sanitaires de l'exposition aux faibles doses de radiations ionisantes a des fins de radioprotection'.DRPH 22005/20 Institut de Radioprotection at de Surete Nucliare. Fontenay aux Roses: IRSN

Ivanov E, Tolochko GV, Shuvaeva LP, et al (1998) Infant leuiemia in Belarus after the Chernobyl accident Eadiat. Env. Biophys. 37 53-5

Kaletsch U, Michaelis J, Burkart W and Grosche B, (1997) 'Infant leukaemia after the Chernobyl Accident' Nature 387, 246.

Petridou E, Trichopoulos N, Dessypris N et al.. (1996) 'Infant leukaemia after in utero exposure to radiation from Chernobyl' BMJ 314, 1200

Savchenko V K, (1995) The Ecology of the Chernobyl Catastrophe: Scientific Outlines of an International Programme of Collaborative Research (Paris: UNESCO).

Stewart A M, Webb J W, Giles B D, Hewitt D, (1956), 'Malignant Disease in Childhood and Diagnostic Irradiation in Utero', Lancet, ii 447.

Wakeford R and Little MP (2003) Risk coefficients for childhood cancer after intrauterine irradiadiation. A review. Int. J.Radiat.Biol 79 293-30

CHAPTER 9
Liquidators Health: a Meta-Analysis

Alexey .V. Yablokov.
Councilor, Russian Academy of Sciences, Moscow.

This review is devoted to the analysis of data concerning the state of health of the participants in the emergency work ("liquidators") following the consequences of the catastrophe in 1986 at the Chernobyl Power Station. Over a period of time, hundreds of thousands of people took part in operations connected with the building of the sarcophagus and in other measures aimed at limiting the consequences of the explosion in the 4th block of the Chernobyl Power Station. The number of liquidators who took part in these operations was approximately 740,000 people between 1986 -1990 (250,000 from Russia, ~360,000 from Ukraine, 130,000 from Belarus).

Undoubtedly, the majority of these people were exposed to additional external and internal radiation. This fact would have adversely affected the state of their health. However, in recent years, specialists connected with nuclear industry have claimed that the health of this group of people, on average, does not differ from that of other population groups in Russia, in Ukraine and in Belarus, and, they claim, sometimes it is actually better. This fact was keenly described in the report of the group of specialists from the "UN Chernobyl forum" "*Health Effects of the Chernobyl Accident and Special Health Care Programmes*", presented under the aegis of IAEA and of WHO.

1. Introductory remarks

While analyzing data regarding the liquidators' health it is necessary to have in view the nature of the initial data. It is well known that, during first years after the catastrophe, doctors were officially prohibited from connecting diseases of the affected population with the radiation. This is why all data concerning the sickness rate within the territories polluted by the Chernobyl radioactive fall-out collected up to 1989 must be considered as false - "probably counterfeited". An obvious example may be taken from the official data referring to the number of cases of leukemia. It is well known that leukemia is the first oncological consequence of radiation. The Russian national register data referring to cases of leukemia among the liquidators, for the first 8 years after the catastrophe, are presented in table 1.

Beginning in 1990, the essential increase in the number of registered cases of leukemia is usually treated as corresponding to peaks of leucosis in the cohort of those irradiated in Hiroshima and in Nagasaki (Tsib, Ivanov, 2000). It is interesting to note that the Japanese and Chernobyl data regarding the increase in leukemia correspond well to the relaxation of the regimes of secrecy.

Table 1 Officially registered leukemia cases for liquidators by date of diagnosis, 1986-1993
(Ivanov et al., 2004, Table 6.6)

Date of diagnosis	Number of cases
1986	1
1987	5
1988	5
1989	3
1990	6
1991	11
1992	9
1993	8

It is well known that the occupying authorities prohibited any research into the consequences of radiation in Japan. That is why the official register only begins in 1950 – four and a half years after the nuclear bombing. In the USSR the prohibition on the diagnosis of radiation-induced diseases was also canceled after 4 years. Therefore, we can assume that the cases of leukemia recorded between 1986-1989 are only a small number of the actual cases of this disease. Officially, "full dosimetric control" of the participants in the work within the zone of the Chernobyl power station took place over only several months (Gerasimova *et al.*, 2001, p. 11). This happened for a number of reasons, some of which are listed below:

- Secrecy;
- Absence of suitable equipment;
- Carelessness of staff and of liquidators;
- Low qualification of staff;
- Deliberate distortion of notes.

An essential distortion of the statistics concerning the illnesses of the liquidators may originally have been connected with the fact that the status of 'liquidator' is not conferred automatically, but only following individual applications. The result is that, today, the status of 'liquidator' refers not only to the persons who were immediately exposed to increased radiation, but essentially to all of the persons who were within the zone of radiation danger in the early period. The official status of 'liquidator' incurs the receipt of a wide spectrum of state benefits (increased pension, habitation, transport, drugs, medical care and others). The conferring of this status is appointed only after numerous formal terms and conditions, the successful execution of which is connected with, not only the presence of objective indexes of worsening of health, but also with a number of subjective factors. Until recent years, the true liquidators, for various reasons, didn't officially register their status as 'liquidator'. Some did not register because of social-psychological reasons (for example: "I am strong, I don't want to go begging"); others for technical (bureaucratic) reasons. Also, it is well known that a number of military personnel who took part in the Chernobyl operations have no official documents confirming their participation in the liquidation work. At the same time, it is well known that among persons who hold this status there are many who obtained it without the correct, objective indexes.

All these factors make it impossible to hold an objective analysis of the statistics relating to the illnesses of the liquidators. Occasionally, these factors deprive the research of any scientific (epidemiological, radiobiological) sense.

On the whole, for an objective evaluation of the connections between the level of illness and the level of additional received radiation, it is clear that the liquidators of 1986-

1987 recieved large doses. With the development of methods for individual biological dosimetry (the determining of additional doses of received radiation by measuring the structure of chromosomes (FISH method) and the calcinated tissue within the tooth (ESR dosimetry)) we have an opportunity to reconstruct and achieve verification of received doses of radiation. The comparison of data gained by these methods show that officially documented doses of radiation may be either top-heavy or understated (Maznik and Vinnykov, 2003).

In spite of the volume of work involved in compiling state registers of the liquidators in Russia, in Ukraine and in Belarus, these data cannot be considered reliable. The official medical authorities acknowledge that the number of Russian liquidators who received doses over $25*10^{-2}$ Gy may be 7 times higher (!) than is shown in the registers (Il'in *et al.*, 1995).

2. General morbidity

Among the people who were sent to Chernobyl, there was no illness– everyone was healthy, and, in general, young. Within five years of the catastrophe, 30% of them were already officially considered sick; within ten years less than 9% of liquidators were considered healthy and within 16 years only about 1-2% left among them were considered to be healthy (Table 2).

Table 2 Dynamics of the liquidator's health condition
(Ivanov et al., 2004; Preebylova et al., 2004)

Year,	after catastrophe	"Sick" persons, %
1986	0	0
1991	5	30
1996	10	90 – 92
2002	16	98 – 99

In Ukraine, 18 years after the catastrophe, the number of sick among the liquidators reached 94,2% (in Kiev – 99,85%, in Sumskaya province – 96.53%, in Donetskaya province – 95,95%) (Chernobyl… 2004).

General morbidity in Ukrainian liquidators increased by 3.5 times over 10 years (Serduk, Bobyleva, 1998). 15 years after the catastrophe this figure increased by three times among Russian liquidators under the age of 30. The annual rates of growth of morbidity among Belarusian liquidators are 2-8 times higher than the control group from the population of Belarus (Antypova et. al., 1997).

As the data in the Russian national register show, the main cause of morbidity among the liquidators in 1996 was diseases of the respiratory organs; of the nervous system; of the sensory organs; of the circulatory system; of musculoskeletal system; of connective tissue and of the endocrine system. In previous years the dynamics for morbidity were different (in 1993; diseases of respiratory apparatus – 15,5 %; of the musculoskeletal system – 14,5%; of digestive system – 14,0%; in 1995; diseases of digestive apparatus – 15,9 %, of the musculoskeletal system and of connective tissue – 14,8 %; of the nervous system – 14,0 %; Problems …, 2002).

Morbidity differs among different age groups of Russian liquidators: in the group of 25-49 year olds mental illnesses and diseases of the nervous system are higher; in the group of 30-59 year olds diseases of the musculoskeletal system, of the endocrine system and of the metabolism are more frequent (Problems …, 2002).

For a number of diseases, the morbidity of the liquidators is notably higher than the morbidity of the population of Russia and Belarus (Table 3, Table 4)

Table 3 Russian liquidators' incidence rate (male, 1996, per 100 000)
(*data from RNMDR by V. Ivanov et al., 2004*)

Disease	RNMDR	Russia	Ratio
Respiratory system	14 215	15 073	0,9
Nervous system and sensory organs	11 041	5 299	2,1
Digestive system	8 613	2 602	3,3
Cardiovascular systems	7 117	1 700	4.2
Musculosceletal system and connective tissue	7 012	3 054	2,3
Endocrine and metabolic	4 637	454	10,2
Mental disorders	2 889	586	4,9
Urogenital system	2 383	3 519	0,7
Skin and subcutaneous tissue	1 552	3 875	0,4
Infection and parasitic	782	3 053	0,3
Neoplasms	713	882	0,8
Blood and blood-forming organs	304	135	2,3
Total	61 687	41 748	1,5

Table 4. Belarus liquidators incidence rate (1995, cases per 100,000 persons)
(Matsko, 1999)

Disease	Liquidators	Belarus population (older than 18)
Thyroid cancer	23.1	7.1
Cataracts	462.8	156.1
Malignant neoplasms (lymphatic and blood-forming organs)	26.1	18.6
Respiration organs diseases	24 781	23 831
Digestion organs diseases	7 784	1 651
Endocrine system diseases, nutritional disorders, metabolism and immunity disorders	3427	518
Blood and blood-forming tissue diseases	304.4	69.4
Mental disorders	3252	1090

According to some data (Gylmanov *et al.*, 2001), general reasons for invalidity of the liquidators are diseases of the cardiovascular system (47,2 %); of the nervous system (19,8 %); diseases of the digestive system (15,4 %) and neoplasm (8,8 %). According to other data (Gerasymova *et al.*, 2001; Problems… 2002) – they are diseases of the nervous system (28,4 %), of the circulatory system (24,0 %) and mental disorders (15,7 %).

Even the data from the national official register (obviously understated) show that the morbidity of Russian liquidators, in a series of indexes, has risen stably over recent years (Table 5).

Table 5 Dynamics of newly detected cases of the Russian male liquidators' diseases (per 100 000) in 1997 – 2001 by RNMDR (*after Ivanov et al., 2004*)

Diseases	1997	1999	2001	% increasing
Neoplasms	1,17	1,47	1,84	158
Blood and blood-forming organs	0,36	0,43	0,56	156
Infections and parasitic	1,14	1,40	1,52	133
Cardiovascular	12,04	14,54	15,52	129

In a wide investigation (3,882 person) the dynamics of general morbidity among liquidators over the 15 years following the catastrophe was observed (Karamalluin *et al.*, 2004). General morbidity among liquidators under 30 years of age at the time of the catastrophe increased by 3 times over 15 years; in the age group of 31-40 year olds, maximum initial morbidity fell in the 8-9 years after the catastrophe.

Among Ukrainian liquidators general sickness increased by 3,5 times during the first ten years (Serduk, Bobyleva, 1998). A noticeable increase in somatic pathology and in oncological diseases (more expressed than in the general population) was observed among Ukrainian liquidators (Golubchykov *et al.*, 2002).

A growing rate was observed also for a number of different illnesses per person (polymorbidity) : before 1991 - 2,8 diseases per person, in 1995 - 3.5, in 1999 – 5,0 (Lyubtchenko and Agal'tzev, 2001).

Common causes of invalidity in the Russian liquidators are diseases of the cardiovascular system (47,2 %), of the nervous system (19,8 %), diseases of the digestive apparatus (15,4 %) and neoplasms (8,8 %), which arrange 91,2% from all causes of invalidity (Gylmanov *et al.*, 2001). Within the list of all diseases, priority belongs to digestive organs pathology, the system of blood circulation, neurological system and sensory organs. These three diseases classes compose 85 - 87 % of all causes of invalidisation in Ukranian liquidators (Bebeshko and Kovalenko, 19994 Prysyazhnyuk *et al.*, 2002). Several years after catastrophe the process of invalidisation became more visible (Table 6).

Table 6 Dynamics of invalidization (per 1,000) among liquidators by dosage groups, 1990-1993
(*based on RSMDR, from Ryabzev, 2002*)

Year	0-5 cGy	5-20 cGy	More than 20 cGy
1990	6,0	10,3	17,3
1991	12,5	21,4	31,1
1992	28,6	50,1	57,6
1993	43,5	74,0	84,7

There are some data relating to a decrease in sexual potency among half of examined male liquidators (Dubivko, Karatay, 2001) and the violation of spermatogenesis among male liquidators (Oganesyan *et al.*, 2002).

3. Mortality

There are 4,136 registered cases of death in the cohort of 52,714 liquidators between 1991-1998, according to official data of the Russian State Register (Ivanov *et al.*, 2002), and during the 15 years following the catastrophe more than 10,000 liquidators died (Gerasymova *et al.*, 2001). The standardized mortality ratio (SMR) varies from 0.78 to 0.88 among this cohort for 4 groups of causes of death:

- from all causes;
- from malignant neoplasms;
- from all causes, except malignant neoplasms;
- from trauma and from poisoning;

Statically this does not differ from the control group of the population. Data regarding the employees of the Kurchatovs Institute show similar results to that of the liquidators (Shikalov at al., 2002). A significant increase in cancer deaths among 66,000 Russian liquidators (the mean calculated dose about 100 mSv) is officially reported for the observation period 1991 – 1998 (Maksioutov, 2002). The ERR coefficient is estimated to be 2.04 Sv^{-1} (95% CI: 0.45 - 4.31).

These data essentially differ from the others. According to data of A.A. Gilmanov *et al.*, (2001) the mortality among male liquidators is higher by 1,4 - 2,3 times than in corresponding age groups within the population. The calculations made by the author of this review showed that the average age of 162 liquidators who died during last 10 years in the town of Tollyaty (Samarskaya province, Russia) was about 46.2 years old (Tymonin, 2005). The average lifespan for 169 liguidators from nuclear industry institutes who died between 1986 – 1990 was 45.5 years (Tukov *et al.*, 2000). In the Kaluga province - National register data, - the average age of death for 84.7 % of liquidators was only 30 - 39 years old (Lushnykov and Lantzov, 1999).

According to the register of liquidators from enterprises in the nuclear field (14,827 men and 2,825 women were examined) the increased level of mortality was among persons with diseases of the circulatory system; of vegeto-vascular and of neuro-circulatory dystonia (Tukov *et al.*, 2002). According to other data (Gylmanov *et al.*, 2001); diseases of the circulatory system (50.9 %); traumas (26.3 %); neoplasms (5.3 %) and diseases of the digestive organs (5.3 %) were the most recent causes of mortality among male liquidators. In 1993 the main causes of mortality for the whole cohort were (Problems... 2002); traumas and poisonings (46 %); diseases of the circulation organs system (29 %) and malignant neoplasms (13 %).

Relative risk of untimely death for this group of liquidators was compiled as follows (Ignatov *et al.*, 2004);

- from malignant neoplasms – 16.3%
- from diseases of the system of hematosis – 25.9%
- from traumas and poisonings – 39,6%
- from other causes– 18,2%

From the Official Russian Inter-agency Expert Council the percentage of mortality among Russain liquidators in the year 2000 was (Khrysanfov and Meskich, 2001);

- 63 % - circulatory pathologies;
- 31% - malignant neoplasms;
- 7% - digestive tract pathologies;

- 5% - lung pathology;
- 5% - traumas and poisoning;
- 3% - tuberculosis.

4. Premature aging

Processes of premature aging are typical for the liquidators. Many liquidators' diseases develop 10-15 years earlier than the average within the population. Among the features of this premature aging are (Javoronkova *et al.*, 2002; Holodova and Zubovsky, 2002; Vartanyan *et al.*, Krasylenko, Eler Ayad, 2002);

- A breach of supreme physical functions;
- An accumulation of diseases typically found in the elderly but inconsistent with the age of the liquidator (10,6 of diagnosis for one liquidator, 2.4 times higher than in control group);
- Degenerative-dystrophic changes in different organs and tissues;
- Accelerated ageing of the brain vessels;
- An imbalance in the antioxidant system.

This research, which lasted 10 years, of a group of 942 invalid liquidators (age 50-70 years old) showed that, in 42 % of cases, cerebral vascular diseases were the main cause of invalidity; in 30 % of cases it was diseases of the cardio vascular system and hypertension. All liquidators had attendant illnesses such as degenerative diseases of the support-motor apparatus, diseases of the alimentary canal and type 2 pancreatic diabetes. These are typical ailments that signify premature ageing (Zubovsky and Malova, 2002).

5. Cancers

By this time, the Russian National Register had accumulated evidence concerning an increase in frequency of some malignant neoplasms among liquidators who worked between 1986-1987 (Ivanov and Tsyb, 2002). By 1996, according to this data, 52 cases of leukemia were found among the liquidators, 47 cases of thyroid cancer and 1,786 cases of other types of cancers. The frequency of leucosis and thyroid cancer was higher than the spontaneous level for corresponding age groups within the population (Ivanov *et al.*, 2001).

All these liquidators' cancer cases, as officially described (Ivanov *et al.*, 2004), agree, within the 95% confidence interval, with rates for the general Russian population. It is noted, by the way, that the estimations of radiation risk are preliminary because the follow-up period is rather short (Ivanov *et al.*, 2004, p. 180).

At the same time, a detailed analysis of concrete groups arrives at quite different conclusions. For instance; the incidence of some malignant neoplasms in female liquidators differs statistically and is higher than the incidence in corresponding age groups within the female population of Russia (Islamova, 2004); the standardized illness relation (SIR) for cancers in all locations is 1.34 (CI 1.17-1.79); for thyroid cancer SIR is 7.85 (CI 3.23-8.77); for cancer of the mammary gland SIR is 1.84 (CI 1.23-2.45).

The most reliable cancer data available for Belarusian liquidators is presented in Table 7.

Table 7 Incidence of cancers among Belarusian male liquidators, 1993 – 2000, per 100 000
(Okeanov et al., 2004)

Site	Liquidators	Control group
All sites*	400,8 ±7,7	361,2±6,4
Colon*	21,6±1,8	16,1±0,6
Urinary bladder*	16,9±1,6	10,4±0,4
Rectum	19,1±1,7	17,9±0,6
Lung	56,9±2,9	53,9 ±1,6
Breast	59,8 ±6,7	57,3 ±0,9
Kidney	16,2±1,6	13,0±0,9

Significant differences.

The incidence rate for leukemia among Ukrainian liquidators is presented in Table 8.

Table 8 Incidence rate for leukemia among Ukrainian liquidators
(Imanaka, 2002)

Year	Incidence rate (per 100 000)	
	Liquidators-1986	Liquidators-1987
1987	13,33± 4,71	-
1988	6,42 ± 3,21	6,32 ± 4,47
1989	14,06 ± 4,69	4,41 ±3,12
1990	14,50 ± 4,59	5,32 ± 3,07
1991	18,13 ± 4,84	7,74 ± 3,46
1992	12,59 ± 3,98	12,02 ± 4,25
Total	13,35 ± 1,80	7,04 ±1,57

The most significant increases in thyroid cancer occurred between 1990-1993 and 1994-1997, among Ukrainian liquidators working in 1986-1987. For liquidators working between 1986-1987, the leukemia and lymphoma incidence rate was significantly elevated between the years 1990-1993 and 1994- 1997. In 1989-1991 there was an observed high risk of leukemia for the liquidators of 1986 in comparison with liquidators of 1987. Russian liquidators from the same period displayed an increase in all types of leukemia, including chronic lymphoid leukemia and chronic myeloid leukemia. In Belorusian liquidators who were exposed between 1986 - 1987 an increase in acute leukemia was seen for 1990 – 1991 (Ledoshchuk and Gudzenko, 2001; Ivanov *et al.*, 1997; Pryayazhnyuk *et al.*, 2002). A statistically significant increase in breast cancer incidence rate in the period 1994-1997 was also observed in female liquidators exposed in 1986-1987 (Pryayazhnyuk *et al.*, 2002).

6. Diseases of the nervous system, sensory organs and psychiatric illnesses.

The level of morbidity from diseases of the nervous system and of the sensory organs among Ukrainian liquidators increased above the average level in the country by 3 times in 1996 (Serduk and Bobyleva, 1998), and among Russian liquidators by 6,4 times in 1995 (About ecological... 2002).

Among liquidators, the number of persons with memory impairments - such as a reduction in the ability to memorize words and images- is unusually high. Such impairment is connected with organic changes within the structure of the frontal lobes of the brain. 80 male liquidators who were examined in 1986 suffering from encephalopathy were found to have a structural / functional inferiority of the frontal lobes of the brain and of the left temporal region, including the cortico-subcortical connections (Antypchuk, 2002, 2003). Among 150 male liquidators (44.5 ± 3 years old) studied, evidence was found of slow forms of the activity and the decreasing of the inter-hemispheric asymmetry {sic}, a decrease in the quality of execution of all cognitive tests, a deterioration in memory and other superior psychological functions (Javoronkova et al., 2002). The average age of the liquidators (male and female) with encephalopathy is 41.2 ± 0.83 years old (Stepanenko, 2003); that is considerably lower than for the population in general.

The clinical and electro-neuro-myographical data show that, among liquidators, the impairment of the peripheral nervous system develops as a result of vegetal-sensing polyneuropathy, which differs from other kinds of poly-neuropathy (Sokolova, 2000).

Research on psychosomatic disturbances among 400 liquidators (24 – 59 years old) shows that they have a marked reduction in functional activity of the CNS, which is conditioned by irreversible damage to the structures of the brain (Antonov et al., 2003, Tsygan et al., 2003).

There is no doubt that psychical disorders are widespread among the liquidators (Nyagu and Loganovsky, 1998; Rumyantseva et al., 2002 and other). 10 years after the catastrophe psychical disorders among Russian liquidators were encountered 4,3 -5 times more frequently than in corresponding groups of the population. A high level of psychiatric disorders among Ukrainian liquidators was observed, especially in 1990-1993 (Nyagu et al., 1999).

Pathological changes in the structure of the brain include atrophy and widening of the ventricles and focal changes (Loganovsky et al., 2003; Nyagu and Loganovsky, 1998). Among the liquidators, indices of neuro-psychic disorders are several times higher then among other population groups within the NPP zone (Nyagu et al., 2002).

Up to 45,9 % of the liquidators have hearing disorders (Zabolotny et al., 2001). Incidence rates for cataracts among Belarusian liquidators are more than twice as high as the general population (Table 9).

Table 9 Incidence rate (per 100 000) of cataracts in Belarus, 1993-1994 (*Goncharova, 2000)*

	Liquidators	Belarus population
1993	281,4	136,2
1994	420,0	146,1

7. Respiratory system

Over time, the incidence of chronic broncho-pulmonary diseases among liquidators increased (Jakushin and Smirnova, 2002; Tseloval'nykova et al., 2003); especially noted are disorders associated with the regulation of breathing caused by a functional decrease in lung capacity and a decrease in elasticity due to impairment of the interstitial structures of the blood-air barrier and the development of peri-alveolar radiation fibrosis (Kuznetsova, 2004). It has been shown that the majority of liquidators that were exposed in 1986-1987 to the inhalation influence of radioactive nuclides have developed progressive breathing disorders (Chykina et al., 2002).

8. Cardiovascular system

The level of morbidity among liquidators with vegetative vascular dystonia, 10 years after the catastrophe, exceeded the average level for Ukraine by 16 times (Serduk and Bobyleva, 1998). 13 years after the catastrophe the level of cardiovascular diseases among Russian liquidators was 4 times higher than in corresponding groups of the population (Bol'shov *et al.*, 2000). The frequency of development and the evidence of disorders of the cardiovascular organs among the liquidators (especially among liquidators exposed in 1986-1987) are especially high compared with the general population. The results of one typical investigation are presented in the table 10.

Table 10 Some characteristics of the cardiovascular system of male liquidators, Voronejskaya province, Russia *(Babkin et al., 2002)*

Index	Liquidators (n=56)	Inhabitants of the polluted territory (n=60)	Control (n=44)
AD – systole	151,9 ± 1,8	129,6± 2,1	126,3± 3,2
AD – diastole	91,5± 1,5	83,2 ± 1,8	82,2±2,2
CHD %	9,1	46,4	33,3
Stroke , %	4,5	16,1	0
The average thickness of the wall of carotid artery, mm	1,71± 0,90	0,81±0,20	0,82± 0,04
Burdened heredity, %	25	25	27,3

From the presented data we can see that the group of liquidators differs from all other groups, which is especially significant bearing in mind that all groups share a similar genetic background.

It was revealed that among 118 liquidators, studied over a period of 15 years, that one third of them had developed ischemic heart disease (Noskov, 2004). In another study, an increase (p<0,05) in the incidence of ischemic heart disease CHD, from 14,6% in 1993 up to 23,0% in 1996, was shown among controlled groups of liquidators (Strukov, 2003), also noted was the syndrome hypodynamia of myocardium and a decrease in the arterial vascular tone of the great circulation (Kovaleva *et al.*, 2004) and an increase of systole arterial pressure (Zabolotny *et al.*, 2001). Among the liquidators of Kishinev (Republic of Moldova) cardiovascular diseases, in recent years, have seen a 3-fold increase (twice that of the control group); 25 % have infiltration of the aorta and 22 % have hypotrophy of the left ventricle (Kirke, 2002). Various disorders of brain circulation have been discovered among the majority of liquidators exposed between 1986-1987 (Romanova, 2001; Bazarov *et al.*,

2001). This tends to occur due to a change in the constructive characteristics of small arteries and arterioles (Troshina, 2004). 10 years after the catastrophe, an examination of 410 liquidators who were exposed in April 1986 has shown that the dynamics of the indices of chronic cerebral vascular pathology, of arterial hypotension, of CHD and of strokes is higher than the indexes of the control group. It is crucial to note these changes among young liquidators as they indicate an increased sensitivity to the factors involved in radiation effects (Kuznetsova *et al.*, 2004).

9. Musculoskeletal system

Pain in the bones and in joints is typical for the vast majority of the liquidators. Approximately 50-57% of examined liquidators have manifestations of osteoporosis and osteopathy. Radiation exposure in the younger age groups often leads to the development of these diseases (Nikitina, 2002; Shkrobot *et al.*, 2003; Kirke, 2002). 12 years after the catastrophe, diffuse osteoporosis was discovered in the jaw structure of 88 % of examined liquidators (Drujynina, 2004).

All liquidators examined at this phase have expressed disorders of vascular blood circulation of the eyes, including crimps {sic}, aneurism and succises {sic}(Rud *et al.*, 2001). General conjunctive and vascular disorders were 10 times higher among liquidators than in the control group (p<0,001).

Typically, these micro-circular disorders among liquidators are a result of damage to the capillary endothelium – the interior lining of the capillaries (Petrova, 2003).

10. Skin

97% of liquidators with psoriasis have developed it after the catastrophe. In all cases, psoriasis is accompanied by functional disorders of the nervous system and impairment of the alimentary canal (Maljuk and Bogdantsova, 2001).

11. Blood and lymphatic System

Disorders connected with the blood and lymphatic system have been noted among the liquidators. These differ considerably from the indices of the control group:

- The average duration of NMR-relaxation of T1 plazma of blood (Popova *et al.*, 2002);
- State of the erythrocytic membranes according to the index of activity of the receptor-lectin reaction (Karpova and Koretskaya, 2003).
- An imbalance of the mean fractions of molecular components in platelets, in erythrocytes and in blood plasma (Zagradskaya, 2002).
- A decrease in the dispersion of the granular multiplier {sic} in the lymphocyte nuclei;
- A decrease of the area of and of the perimeter distance of the ingranular zone and a frequent increase in irregularity within this zone (Akulich, 2002);
- Change in the quantity of leukocytes, erythrocytes and lymphocytes in the peripheral blood (Tukov *et al.*, 2002).

Even after ten years, lymphopoiesis among liquidators remains disturbed (Table 11):

Table 11 The dynamics of the correlation of types (%) of lymphopoiesis among Russian liquidators (*by Karamullin et al., 2004)*

	Types of the lymphopoiesis		
	Quasi-normal	Hyper-reactivation	Hypo-reproducer
Liquidators about 5 years after the catastrophe	32	55	13
Liquidators 5-9 years after the catastrophe	38	0	62
Liquidators 10-15 years after the catastrophe	60	17	23
Healthy group (control)	**76**	**12**	**12**

12. Diseases of the digestive system and other internal organs

Diseases of the digestive system among Belarusian liquidators are registered 3,5 - 4 times more frequently than among the adult population of the country (Antypova *et al.*, 1997). A controlled study of 118 liquidators, over 15 years revealed (Noskov, 2004):
- Changes in the liver - 40,6 % ;
- Changes in the structure of the pancreas - 60,2 %;
- Thickening of the gall-bladder wall – 29 %;
- Urolithiasis - 25 %;
- Changes in the thyroid gland - 60,2 %.

Structural changes of the pancreas increase with time (Table 12).

Table 12 Dynamics of change (% of examined) within the pancreas among Ukranian male-liquidators (*Kamarenko et al., 2002; Komarenko and Polyakov, 2003)*

	1987-1991	1996-2001
Thickening of the gland	31	67
Increasing of the echogenicity of the tissue	54	81
Change of the structure	14	32
Change of the contour	7	26
Change of the capsule	6	14
Widening of the Wirsung's duct	4	10
All changes of the echostructure	37,6 (1987)	87,4 (2002)

13-15 years after the catastrophe, liquidators, in comparison with control groups, have (Pimenov, 2001; Drujynina, 2004);
- Intense development of caries;
- Pathologic dental abrasion;

- Diseases of the peridontium and of the mucous membrane within the mouth

In 1996, incidents of stomach ulcers among Ukrainian liquidators were 3,5 times higher than the average for the Ukraine (Serduk and Bobyleva, 1998).

13. Disturbances of the hormone status

Morbidity of the endocrine system among Russian liquidators was 10 times higher in the 13 years after the catastrophe than in corresponding groups of the population (Bol'shov *et al.*, 2000). Deep reorganization of pituitary regulation and changes in hormone production by the endocrine glands was revealed among liquidators (Drygina, 2002). 22 % of examined male-liquidators have increased amounts of prolactin (the hormone of the pituitary), which is typical only for young women (Strukov, 2003).

During 1990-1994 the risk of diseases of the thyroid gland was 8,9 times higher among Belarusian liquidators than among the adult population of Belarus and by 1996 diseases of the thyroid gland among liquidators were 11, 9 times more common than among the adult population of Belarus (Antypova *et al.*, 1997). In a large group of Russian male liquidators exposed in 1986-1987 (1,752 persons), an essential increase was revealed in the number of cases of structural changes in the tissue of the thyroid gland 6-7 years after the catastrophe (Table 13).

Table 13 Dynamics (number of cases in %) of structural changes of the tissue of the thyroid gland of Ukrainian male liquidators (*according Lyashko et al., 2000*).

Feature	1992 г.	1994 г.
Nodal formation	13,5	19,7
Hyperplasia	3,5	10,6
Thyroiditis	0,1	1,9

Morbidity from chronic thyroiditis in Ukrainian liquidators between the mean for the period 1992 -1995 to that in 2001 increased by 154 % (Moskalenko, 2003). The majority of 500 liquidators examined during the first years after the catastrophe had essential disturbances to the pituitary suprarenal system. In 6 years there was a re- normalization of the study indexes for the resting state, but this re-normalisation did not occur for condition of functional load (Mitryaeva, 1996).

14. The growth in the number of genetic abnormalities – mutations

Numerous studies shortly after the catastrophe showed the presence of radiogenic chromosome mutations in the lymphocytes of liquidators, a number of which was correlated with a received dose of radiation (Shevchenko *et al.*, 1995; Svirnovsky *et al.*, 1998; Begenar, 1999; Shykalov *et al.*, 2002). During the first years after the catastrophe, the number of unstable (dicentric and ring chromosomes) and also of stable aberrations of chromosomes (translocation, insertion) essentially increased (Oganesyan *et al.*, 2002; Demyna *et al.*, 2002; Maznik, 2003). Over time, rates of elimination of aberrations of the chromosome type and of genome irregularities among liquidators who were exposed to higher radiation decreased (Maznik and Vinnykov, 1997). 10-15 years after the catastrophe, the level of stable aberrations in the lymphocytes of the peripheral blood of the liquidators steadily increased (Mel'nikov *et al.*, 1998; Pilinskaya, 1999; Pilinskaya *et al.*, 2003).

Among liquidators exposed in 1986, the highest level of genetic damage was observed in the builders of the sarcophagus and the dosimetrists (Table 14).

Table 14 Cytogenetic characters of the liquidators of 1986 within 3 months
(Shevchenko, Snegireva, 1999)

Group	Persons, n	Cells scored, n	Chromosome aberrations, n	Dicentrics + centric rings
Chernobyl Nuclear Power Plant staff	83	6015	23.7 ± 2.0*	5.8 ± 1.0*
Builders of the "Sarcophagus"	71	4937	32.4 ± 2.5*	4.4 ± 0.9*
Dosimetrists	23	1641	31.1 ± 4.3*	4.8 ± 1.7*
Drivers	60	5300	14.7 ± 1.7*	3.2 ± 0.8*
Physicians	37	2590	13.1 ± 2.3*	2.7 ± 1.0*
Pripyat population	35	2593	14.3 ± 2.4*	1.9 ± 0.8*
Control	19	3605	1.9 ± 0.7*	0 - -

* $p < 0.05$;

A cytogenetic study, which was performed 8 - 9 years after the catastrophe, shows that in this group of liquidators the frequency of cells with translocations exceeds that of the control group by nearly 4-fold (Shevchenko, Snegireva, 1999). The number of chromosome aberrations expressed a clear peak in 1992 - 1993 (Table 15).

Table 15 Dynamics of cytogenetic characters (Mean $\pm$ SEM, 10^{-3}) of lymphocytes in the liquidators 1990 – 1995 *(Shevchenko, Snegireva, 1999)*

	Persons, n	cells scored, n	chromosome aberrations, n	Dicentrics + centric rings	Cdr + ace*
1990	23	4268	14.9 ± 1.9*	1.0 ± 0.5*	5.3 ± 1.1
1991	110	20077	19.7 ± 1.0*	0.9 ± 0.2*	6.8 ± 0.6*
1992	136	32000	31.8 ± 1.0*	1.4 ± 0.2*	9.0 ± 0.5*
1993	75	18581	34.8 ± 1.4*	0.9 ± 0.2*	11.9 ± 0.8*
1994	60	181793	1.8 ± 1.3*	1.8 ± 0.3*	10.3 ± 0.8*
1995	41	12160	18.8 ± 1.2*	0.4 ± 0.2	7.3 ± 0.8*
Control	82	26849	10.5 ± 0.6	0.2 ± 0.1	3.9 ± 0.4

* *$p < 0.05$;*
* *Cdr - cells containing dicentrics and/or centric rings; ace - acentrics.*

15. Disturbances of the immune system

The immune system of the liquidators is in an unstable condition, resulting in a state of immunodeficiency. Many liquidators, 10-15 years after the catastrophe, have a deviation in the quantitative indexes of cellular and humeral immunity and changes in immune status (Melnov *et al.*, 2003; Matvienko *et al.*, 1997; Shubnik, 2002; Gajeeva *et al.*, 2001; Novykova, 2003; Soloshenko, 2002; Korobko *et al.*, 1996; Potapnev *et al.*, 1998; Bazyka *et al.*, 2002; Maluk and Bogdantsova, 2001; Grebnyuk *et al.*, 1999).
These changes appear in:
Changes in the correlation of the sub-population of T-lymphocytes,
    T-helpers and T-supressors;
A general decrease in the amounts of T- and of B-lymphocytes;
A decrease in the level of serum immune globulin of A,G and M
    classes;
Irregularities in cytokine production;
Activation of the neutrophilic granulocytes.

A frequency of morbidity due to allergic diseases such as rhinitis (increased by 6-17 times) and hives (increased by 4-15 times) has occurred in liquidators from Obninsk city (Russia) 7-9 years after the catastrophe in comparison with the city population (Tataurstchykova *et al.*, 1996). 4 years after the catastrophe only 17 % of examined liquidators had restored levels of neuropeptide dermophin, and more then 50 % had decreased levels of other neuropeptides (of leu- and of methionine-enkephalin) (Sushkevich *et al.*, 1995).
    The majority of 400 examined Ukrainian liquidators have changes within the ultrastructure of neurophils (destruction of the contents, hyper-segmentation of the nucleus, anomalous polymorphous growths and others) and in the lymphocytes (increased sinuosity in the contour of the membrane, a segregation of the chromatin and of structural components with the nucleolus) (Zak *et al.*, 1996).

17. Health of the liquidators' children

There are more and more data concerning the abrupt decline in the health of the liquidators' children. Ukrainian liquidators' children have higher morbidity of the digestive organs, of respiratory apparatus, of the nervous system and of the endocrine system. They are subject to a number of congenital malformations, hereditary conditions and increased frequency of infectious diseases (Ponomarenko *et al.*, 2002).
    In areas where detailed studies of the families of liquidators took place (for example in Razanskaya province; Lyaginskaya *et al.*,2002), the following was shown:
- An increase in frequency of sick newborns;
- An increase in frequency of congenial malformations;
- An increase in frequency of births of children with a birth weight lower than 2,500 grammes.
- A delay in pre-natal development;
- A higher frequency of general sickness.

Such illnesses are more common among the children of liquidators exposed between 1986-1987 (Table 16).

Table 16 The health of Russian liquidators' children
(after Lyaginskaya et al., 2002)

	Father received 5 and more cSv	Father received less 5 cSv	All children in Ryazan' procvince
Coefficient of the prenatal development delay	3	2	1
Morbidity	Highest	Intermediate	Low
Congenital malformation number	Highest	Intermediate	Low
Unhealthy newborns	Many newborn	Intermediate	A few

In studies of morbidity in liquidator's children there is a prevalence of chronic disorders of the ENT organs, deviations of red blood cells, fundamental abnormalities of the nervous system, plural caries, chronic gingivitis and anomalies of the dental-jaw system (Marapova and Hytrov, 2001).

It is also interesting that examined children have an increased number of chromosome aberrations in the somatic cells. The number of deletions, of inversions of the rings, of isochromatid and single fragments, of breaks and of cases of polyploidy has increased (Ibragymova, 2003).

Even according to the evidently understated data from the Russian National Register, among liquidators' children, less than 30 % are listed in the group regarded as "healthy" (Table 17).

Table 17 Health conditions of the liquidators' children
(V. Ivanov et al. 2004)

Group	Liquidators (%)	Their children (%)
"Healthy"	8,6	27,8
"Need further examination"	24,0	54,4
"Sick"	67,4	17,7

18. Conclusion

This review is based on material that was published largely between 2000 - 2004 and discusses data concerning the heath of liquidators 10-15 years after the catastrophe. The total number of publications addressing this problem includes more than 1,500 articles and other papers and to summarize this volume of research would require not an article but several monographs. However, the data from this review is enough to form a conclusion: The health of persons who took part in measures to minimize the consequences of the Chernobyl accident, and who received additional doses radiation, is disastrous. Liquidators die much earlier, fall ill more frequently and remain ill longer (with many diseases) than people from corresponding age groups throughout Russia, Belarus and Ukraine.

This happens in spite of regular, additional and intensive medical care, including drugs and sanatorium / spa treatments. Thanks to more careful medical examinations we can definitely say that the consequences of radiation in small doses cannot be reduced to "radio-phobia", or to "victimization", or to social-economic problems, as is widely claimed by specialists connected with the nuclear industry.

Material concerning the health of liquidators shows that some structures of the nervous system (including the central nervous system) are especially damaged by low-level radiation. Different epithelial cells (including the lining of the capillaries) and the immune system are also acutely vulnerable

Data relating to liquidators' health show that the official falsification of medical data between 1986-1989 exclude the possibility of an accurate estimation of the full consequences of the Chernobyl catastrophe. For example the "UN Chernobyl Forum" states that 50 liquidators have died because of acute radiation sickness, about 300 liquidators have died from cancers and less than 5,000 additional Chernobyl-related cancer deaths are predicted for the future. However, the data considered here show that these figures must be increased by a level of 2-3 orders of magnitude (i.e. up from 100 to 1,000 times higher).

In pursuance of improving the efficiency of treatment for the liquidators it would be expedient to obtain data from individual ESR (dosimetry of tooth enamel) and the FISH (Fluorescence *In Situ* Hybridisation) method, in order to determine the damage to the DNA, which would provide a better retrospective measure of dose than the official records, which are clearly incorrect.

Literature

About the ecological factors of the worsening of the demographical situation. 2002. «Ecological Security of Russia». Material of Interagency Commiss. On Ecolog. Security, RF Security Council (September 1995 – April 2002), 4, Moscow, pp. 211-225 (in Russian).

Akulich N.V. 2003. Epigenotype of the lymphocytes among persons exposed the ionizing impact. Selected Sci. Papers, Mogilev' State Univ. name A.A. Kuleshov, Mogilev, pp.204 -207 (in Russian).

Alymov N.I., Pavlov A.Yu. Sedunov S.G., Gorshenin A.V., Popovich V.I., Loskutova N.D., Belobrovkin E.A. 2004. The estimation of the condition of the immune system of the organism of the persons living on the territory exposed to the radiation impact because of the catastrophe on Chernobyl NPP. «Med.-biol. Problems of anti-radiation and anti-chemical protection". Collect. Of Papers Russian Sci. Confer., Sankt-Petersburg, May 20-21, 2004, Sankt-Peterburg, pp. 45 - 46. (in Russian).

Antypova S.I., Korjunov V.M, Polyakov S.M., Furmanova V.B. 1997. Problems of the health state of the participants of the liquidation of the consequences of the Chernobyl's catastrophe. «Med.-biol. Effects and ways to overcome the consequences of the Chernobyl accident», Collect. of Sci. Papers, devout. 10th anniversary of Chernobyl' accident. Minsk – Vitebsk, p. 3 (in Russian).

Antypova S.I., Korjunov V.M., Suvorova I.V. 1997a. The tendencies of liquidator's morbidity with chronic non-specific diseases. «Actual problems of the med. rehabilitation of the suffers after Chernobyl' catastrophe», Materials Sci.- Pract. Conf., devout. By 10th anniversary of Republican Hospital of Radiation Medic. (Minsk, 30 June, 1997), Minsk, pp. 59 – 60 (in Russian).

Antypchuk K.Yu. 2003. Memory state of the persons who have had the acute radiation sickness because of the catastrophe of Chernobyl's NPP in the remote period. Ukrainian Radiol. Jour., vol. 11, № 1, pp. 68 -72 (in Ukrainian).

Antonov M.M., Vasil'eva N.A., Dudarenko S.V., Rozanov M.Yu., Tsigan V.N. 2003. Mechanisms of forming of psychosomatic irregulations under the influence of small dozes of ionizing radiation. New Med. Thechnol. Herald, vol. 10, № 4, pp. 52 – 54 (in Russian).

Arynchina N. T., Milkamanovich V.K. 1992. The Comparison characteristic of the 24-hour monitoring data of heart arrhythmias among invalids with the ischemic heart disease living on the territories polluted by the radioactive nuclides and in clean regions of Belarus. Abstracts, Jubille Conf. 125th Belarus Sci. Therapeutic Soc., Minsk, 22-23 December, 1992, Minsk, pp. 75 –76 (in Russian).

Babkin A.P., Choporov O.N., Kuralesin N.A., 2002. The features of the diseases of the cardiovascular system of liquidators of the consequences of the Chernobyl's catastrophe and of the persons who live on polluted territories. Labor and Industr.-Ecolog. Med., № 7, pp. 22 – 25 (in Russian).

Baeva E.V., Sokolenko V.L., 1998. Expression of T-cellular superficial marker by the lymphocytes of persons exposed to the small dozes of radiation. Immunology, № 3, pp. 56 – 59 (in Russian).

Bazarov V.G., Belyakova I.A., Savchuk L.A., 2001. Cerebral dynamic under the experimental vestibular stimulation among liquidators of the catastrophe consequences. Jour. Ear, Nose and Throat Illness, № 4, pp. 1 – 5 (in Ukrainian).

Bazyka D., Chumak A., Byelyaeva N. 2003. Immune cells in Chernobyl radiation workers exposed to low dose irradiation. Int. J. Low Radiation, vol. 1, pp. 63 -75.

Bebeshko V.G., Kovalenko A.N. (Eds.). 1999. Medical Consequences of Chernobyl nuclear power station accident. Book 2. Clinical aspects of the Chernobyl catastrophe. Kiev, "MEDECOL" interdisciplinary Scientific and Research Centre BIO-ECOS, 399 p. (In Russian).

Begenar V.F. 1999. Immune hematological and cytogenetic aspects of the impact of small dozes of radiation on women organism. Russ. Associate Obstetric - Gynecol. Herald, № 1, pp. 33 – 36 (in Russian).

Bero M.P., 1999. The distribancies of sexual health of male liquidators. Jour. Psych. Med. Psychology, № 1(5), pp. 64 – 68 (in Russian).

Bol'shov L.A., Aruthyunyan P.V., Linge I.I., Barkhudarov R.M., Osypyants I.A., Gerasymova N.V., Blinov B.K., Marchenko T.A., Zyborov A.M., 1999. Chernobyl's catastrophe. Results and problems of overcoming of its consequences in Russia 1986-1999. (www. ibrae. ac. Ru /russian/chernobyl/nat_rep_99/13let_text.html) (in Russian).

Chernobyl: medical consequences (in 18 years after). 2004. LigaBisnessinform, 22 april

Chekyna S.Yu., Pashkova T.L., Kopylev I.D., Chernyaev A.L., Samsonova M.V., Aysanov Z.R., Chuchalin A.G. 2002. Functional state of the respiratory system of the liquidators of the Chernobyl's catastrophe: the results of seven years observation. Pulmonology, № 4, pp. 66 -71(in Russian).

Chuchalin A.G. 1996. Pathology of respiratory organs among the liquidators of the consequences of the Chernobyl's catastrophe. Therapeutic archives, vol.68, № 3, pp. 5-7 (in Russian).

Gajeeva T.P., Tchekotova E.V., Krotkova M.V. 2001. The state of the immunity of male liquidators after the Chernobyl's catastrophe. 11th Intern. Symp. Bioindicators. "Modern problems of Bioindication and Biomonitoring", Syktuyvkar, 17-21 September, 2001. Syktuyvkar, p. 31 (in Russian).

Gerasymova N.V., Blinov B.K., Marchenko T.A., Zyborov A.M., Onischenko G.G., Ivanov S.I., Permynova G.S., Goncharik N.V., Kurganov A.A., Bol'shov L.A., Aruthyunyan P.V., Linge I.I., Barkhudarov R.M., Belyaev S.T., Tsib A.F.,

Ivanov V.K., Alexahin R.M., Il'in L.A., Izrael Yu.A. 2001. Chernobyl's catastrophe. Results and problems of overcoming of its consequences in Russia 1986-2001. Russian National Report , RF Ministry of Emergency Situation, Moscow, 39 p. (www.ibrae.ac.ru/russian/nat_rep2001.html) (in Russian).

Golubchykov M.V., Mikhenko Yu. A., Babynets A.T. 2002. The changes in the state of health of the population of Ukraine in post-catastrophe period: Reports 5-th Anual sci. pract. Conference «To 21 century with safety nuclear technology», Slavytich, 12-14 April, 2001, Scientific and technological aspect of the Chernobyl, № 4, pp. 579-581 (in Ukrainian).

Goncharova R.I. 2000. Remote consequences of the Chernobyl Disaster: Assesnment After 13 Years. In: E.B. Burlakova (Ed.). Low Doses Radiation: Are they Dangerous? NOVA Sci. Publ., pp. 289 – 314.

Gylmanov A.A., Molokovich N.I., Sadykova F.H. 2001. Health state of liquidators' children. Diagnostics, treatment and rehabilitation. Materials Intern. Interdisciplinary Sci.- Practical Conf., devoted by 15th anniversary of the Chernobyl' catastrophe, Kazan', 25 - 26 April, 2001. Kazan', pp. 25 – 26 (in Russian).

Demyna E.A., Klyushin D.A., Petunin Yu.I. 2002. Cytogenetic and carcinogenic effects of small dozes on liquidators of the consequences of the Chernobyl's catastrophe. Abstracts, 3rd Intern. Symp. "Mechanism of the Ultra-Law doses action", Moscow, 3 - 6 December, 2002, Moscow, p. 71 (in Russian).

Dregyna L.B. 2002. Clinic-laboratory estimations of the state of adaptive-regulation systems of liquidators in remote times. Doc. Thesis, Biol. Sci., All-Russian Center Emergency and Radiation Medic., Sank- Petersburg, 25 p. (in Russian).

Drujynina I.V. 2004. The condition of the bone tissue of jaws of persons who took part in the liquidation of the consequences of the Chernobyl's catastrophe. Materials Interregion. InterUniversities Sci. Students Conf., Perm', 5-7 April, 2004. vol. 1, Perm' - Izhevsk, pp. 53 -54 (in Russian).

Dubivko G.F. Karatay Sh.S. 2001. Dependence of the sexual function of a man from the stressors and radioactive impacts. Diagnosis, cure and rehabilitation suffering in emergency cases. Materials Intern. Interdisciplinary Sci.-practical Conf., devoted by 15th Anniversary of the Chernobyl' Catastrophe, Kazan', 25 - 26 April, 2001 , Kazan', pp. 113 -117 (in Russian).

Grebenyuk A.N., Bejenar' A.F., Antushevitch A.E., Lutov R.V. 1999. About estimation of the immune statue of women after radiation and chemical factors impact. Military.-Med. Jorn., № 11, pp. 49 – 54 (in Russian).

Ibragymova A.I. 2003. Clinic data about the genotoxic influence of the ionizing radiation. Russ. Herald of Perinatology and Pediatric, vol. 48, № 6, pp. 51 – 55 (in Russian).

Imanaka T. (Ed.). 2002. Recent Research Activities about the Chernobyl NPP Accident in Belarus, Ukraine and Russia. Recent Research Activities about the Chernobyl Accident in Belarus, Ukraine and Russia. Kyoto University Research Reactor Institute. Kurri-Kr-79, July

Islamova A.R. 2004. Morbidity with the malignant neoplasms of female liquidators. Doc. Thesis, Med. Sci., Obninsk, Med. Radiolog. Sci. Center, 19 p. (in Russian).

Ivanov V.K., Matveenko E.G., Birjukov A.P. 1996. Analysis of a new revealed sickness of the liquidators of Kalugskaya province. «Chernobyl Legacy», Mater. Sci.-Practical Conf. "Med.-Psycholog., Radioecological and Social-Economical aspects of liquidation of consequences oa Chernobyl Accident in Kaluga Province", Kaluga - Obninsk, 2, pp. 233 – 234 (in Russian).

Ivanov, E.P. et al. 1997. Chernobyl registry and hematological surveillance of Belarus liquidators. p. 171-192 in: Sixth Symposium on Chernobyl-Related Health Effects,

Tokyo, 10-11 December.

Ivanov V.K., Tsyb A.F., Nilova E.V. 1997. Cancer risks in the Kaluga oblast of the Russian Federation 10 years after the Chernobyl accident. Radiat. Environ. Biophys. Vol. 36, pp. 161-167.

Ivanov V.K., Rastopchin E.M., Gorsky A.L. 1998. Cancer incidence among liquidators of the Chernobyl accident: solid tumors, 1986-1995. Health Phys., vol. 74, pp. 309 - 315.

Ivanov V.K., Tsib A.F., Ivanov S.I., 1999. Liquidators of the Chernobyl's catastrophe: radiation-epidemiologic analysis of medical consequences. Moscow, Publ. "Galanis", 312 p. (in Russian).

Ivanov V., Tsyb A., Ivanov S., Pokrovsky V. 2004. Medical Radiological Consequences of the Chernobyl Catastrophe in Russia. Estimation of Radiation Risk. St. Petersburg, Publ. "Science", 388 p. (in Russian).

Il'in L.A., Kryuchkov V.P., Osanov D.P., Pavlov D.A., 1995. Levels of the radiation of the liquidators in 1986-1987 and the verification of dosimetric data. Radiat. Biol. Radioecology, vol. 35, 6, pp. 803 – 827 (in Russian).

Javoronkova L.A., Rygov B.N., Barmakhova A.B., Kholodnova N.B., 2002. Features of the disturbances of EEG and of cognitive functions after the impact of the radiation. Russ. Acad. Sci. Reports, vol. 386, № 3, pp. 418 – 422 (in Russian).

Kamarenko D.I., Glukhen'kiy E.V., Polyakov O.B. 2002. Structural changes of the pancreas of the liquidators of the Chernobyl's catastrophe, which were revealed by the prospective sonographic data in the remote period. Ukran. J. Hematology Transfusiol., № 5, pp. 28 – 32 (in Ukrainian).

Karamullin M.A., Sosukin A.E., Shutko A.N., Nedoborsky K.V., Yazenok A.V., Ekymova L.P., Grischuk A.V., Babhak A.V. 2004. Importance of the factor of the doze of the radiation for the forming of the sickness in the remote period among the participants of the liquidation of the consequences of the Chernobyl's catastrophe according of their age. Abstracts, "Actual Questions of Radiation Hygiene", Sci.-Practical Conf., Sankt-Peterburgh, 21 - 25 June, 2004, Sankt-Peterburg, pp. 170 – 171 (in Russian).

Karpova I.S., Koretskaya N.V., 2003. The influence of the character of the radiation doze on the activity of the receptor-lectin reaction at liquidators of the Chernobyl's catastrophe. Biopolymers and Cell, vol. 19, № 2, pp. 133 -139 (in Ukrainian).

Kirke L. 2002. Early development of some diseases among the liquidators of the Chernobyl's catastrophe. Reports, 7th Intern. Sci.-Practical Conf. "Aging Pacient. Quality of Life", Moscow, 1 - 3 October, 2002, Hospital Gerontology, vol. 8, № 8, p. 83 (in Russian).

Kharchenko V.P., Zubovsky G.A., Kholodnova N.B. 1995. Changes in the brains among the liquidators of the consequences of the Chernobyl's catastrophe. Herald of Roentgenlogy and Radiology, № 1, pp. 11 – 14 (in Russian).

Kholodova N.B., Zubovsky G.A. 2002. Polymorbidity like a syndrome of the premature aging in the remote period after the irradiation with the small dozes. Reports, 7th Intern. Sci.-Practical Conf. "Aging Pacient. Quality of Life", Moscow, 1 - 3 October, 2002, Hospital Gerontology, vol. 8, № 8, p. 86 (in Russian).

Klimenko D.I., Snysar I.A., Samophalova E.G. 1996. Immunologic reactivity and functional state of hearing and of vestibular analyzer of liquidators of the Chernobyl's catastrophe. Remote consequences of irradiation for immune and haemopoetic systems. Abstarcts, Sci.-Pract. Conf., 7-10 May, 1996, Kiev, pp. 29 – 30 (in Ukrainian).

Komarenko D.I., Polyakov O.B. 2003. Post-radiational pancreatopathy: remote consequences of the ionizing radiation. Gastro-Enterology Herald, № 1, pp. 31 -35 (in Ukrainian).

Korobko I.V., Korit'ko S.S., Blet'ko T.V., Korbut I.I. 1996. Features of functioning of the interferon system among the liquidators. Immunology, № 1, pp. 56 – 58 (in Russian).

Krasylenko E.P., Eler Ayad M.S. 2002. Age features of the intersystem correlation of the cerebral hemodynamics among persons with high genetic or ecological risks of the development of the cerebro-vascular pathology. Aging and Longevity Problems, vol. 11, № 4, pp. 405 – 416 (in Russian).

Kuznetsova S.M., Krasylenko E.P., Kuznetsov V.V. 2004. Vascular diseases of the brain and cerebral blood circulation of the participants of the liquidation of the Chernobyl's catastrophe. Hospital Gerontology, vol. 10, № 8, pp. 18 – 28 (in Russian).

Law. 1996. Federal law "about the introduction of changes and of additions into the Law of Russian Federation "About the social protection of citizens exposed to the influence of the radiation because of the catastrophe on Chernobyl NPP", December 11, 1996.

Leberman A.N. 2003. Radiation and reproductive health. Sankt-Peterburg, Publ. "New Century", 225 p. (in Russian).

Ledoshchuk B.A., Gudzenko N.A. 2001. Haemoblastosis in liquidators: 15 years after Chernobyl catastrophe. Main results and perspective. In: International Conference "Fifteen Years after the Chernobyl Accident. Lessons Learned", Kiev, Chornobylinterinform. pp. 3-10 (in Russian).

Loganovsky K. Medical consequences of the Chernobyl's catastrophe: what do we know after 19 years? (www.chornobyl.net/ru/261/prn/) (in Russian).

Loganovsky K. N. 1999. Clinic-medical aspects of the psychiatric consequences of the Chernobyl's catastrophe. Social and Medical Psychiatry, vol. 9, 1, pp. 5 – 17 (in Russian).

Loganovsky K.M., Kovalenko O.M., Yur'ev K.L., Bomko M.O., Antypchuk K.Yu., Denisjuk N.V., Zdorenko L.l, Rossokha A.P., Chornij A.G., Dubrovena G.V. 2003. Verification of the organic affection of the brain in the remote period of the acute radiation sickness. Ukrainian Medical Annual Report, № 6, pp. 70 – 78 (In Ukrainian).

Lushnykov E.F., Lantsov S.I. 1999. Mortality of liquidators in Kalujskaya province in 10 years after the catastrophe on Chernobyl's NPP. Med. Radiol. and Radiat. Safety, vol. 44, № 2, pp. 36 – 44 (in Russian).

Lyaginskaya A.M., Osypov V.A., Smirnova O.V., Isichenko I.B., Romanova S.V. 2002. Function of reproduction of the liquidators of the consequences of the Chernobyl's accident. Med. Radiol. and Radiat. Safety, vol. 47, № 1, pp. 5 – 10 (in Russian).

Lyashko L.I., Tsib A.F., Sushkevich G.N. 2000. Radioactive nuclide's methods in the diagnostics of the thyroid gland diseases among the liquidators of the Chernobyl's catastrophe. Intern. Conf. "Modern Problems of Nuclear Medicine and Radio-pharmacology", 2nd Congr. Russ. Soc. Nuclear Medicine, Obninsk, 23 - 27 October, 2000, Obninsk, pp. 95 – 96 (in Russian).

Maksioutov M.M. 2002. Radiation Epidemiological Studies in Russian National Medical And Dosimetric Registry: Estimation of Cancer and Non-Cancer Consequences Observed among Chernobyl Liquidators. In: Imanaka T. (Ed.). Recent Research Activities about the Chernobyl Accident in Belarus, Ukraine and Russia. Kyoto University, Research Reactor Institute, Kyoto. Pp. 168 – 187.

Maljuk E.S., Bogdantsova E.N. 2001. The features of the appearance and of the clinical course of psoriasis of the liquidators of the consequences of the Chernobyl's NPP. Collect. Of Sci. Papers «185 years Krasnodar Region. Med. Hospital, named Prof. S.V. Otchapovsky», Krasnodar, p. 134 (in Russian).

Marapova L.A., Hytrov B.Yu. 2001. The stomatologic status of liquidator's children. «Diagnostic, cure and rehabilitation of sufferers during emergency operations», Materials Intern. Interdisciplinary Sci.-Pract. Conf., devoted by 15th anniversary of the Chernobyl catastrophe". Kazan', 25 - 26 April, 2001, Kazan', pp. 193 -195 (in Russian).

Matsko M.P. 1999. Current state of Epidemiological Studies in Belarus about Chernobyl Suffers. In: Imanaka T. (Ed.). Research Activities about the Radiological Consequences the Chernobyl NPS Accident and Social Activity the Suffers by the Accident. Kyoto University, Research Reactor Institute, Kyoto, pp.127 -138.

Matvienko V.N., Javoronok S.V., Sachek M.M. 1997. Flowing cytometric analysis of subpopulations of lymphocytes of the peripheral blood of the liquidators. «Med.-Biol. Effects and ways to overcome of the consequences of the Chernobyl NPP Accident», Collect. of Sci. Papers, devoted by 10th anniversary of Chernobyl NPP' Accident. Minsk- Vitebsk, pp. 34 – 36 (in Russian).

Maznik N.A., Vinnykov V.A. 1997. The dynamics of the Cytogenetic effects in the lymphocytes of the peripheral blood of liquidators of the Chernobyl's catastrophe. Genetics and Cytology, vol. 31, № 6, pp. 41 – 46 (in Russian).

Maznik N.A., Vinnykov V.A., Maznik V.S. 2003. The estimation of the distribution of individual dozes of the radiation of the liquidators of the consequences of the Chernobyl's catastrophe, according to data of the cytogenetic analysis.Radiat. Biol., Radioecology, vol. 43, № 4, pp. 412 – 419 (in Russian).

Medical consequences of the Chernobyl's catastrophe. 1995. The results of the IPHECA Pilot Projects and National Programmes. Sci. Report, WHO, Geneva, 560 p.

Mel'nikov S.B., Koryt'ko S.S., Greschenko M.V. 1998. The dynamic of the cytogenetic status of the liquidators. Public Health, № 2, pp. 21 – 23 (in Russian).

Mel'nov S.B., Koryt'ko S.S., Aderyho K.N., Kondrachuk A.N., Shymanets T.V., Nikonovich S.N. 2003. The estimation of the immunological status of liquidators of 1986-87, in remote periods after the participating in catastrophe works. Immuno-pathology, Allergology and Infectology, № 4, pp. 35 – 41 (in Russian).

Mitryaeva N.A. 1996. The hypothalamo-pituitary-adrenal axis of the liquidators of the consequences of the Chernobyl's NPP. Med. Radiology and Radiol. Safety, vol. 41, № 3, pp. 19 – 23 (in Russian).

Mitjunin A. 2005. National consequences of the liquidations of the Chernobyl's catastrophe. Atomic Startegy at XXI Century, № 1, 22 p. (in Russian).

Nekytina N.V. 2002. Study of the mineral density of the bone and of the intensity of the exchange of the bone tissue of the liquidators of the consequences of the Chernobyl's NPP. Abstarcts, 6th Region. Conf. Junior Researchers of Volgograd Province, Volgograd, 13 -16 November, 2001 , Volgograd, pp. 87 – 88 (in Russian).

Noskov A.I. 2004. Visceral organs' pathologies in Chernobyl' NPP liquidators under 15th years observations. Materials Sci. - Pract. Conf. with Intern. Participation and Workshop for Junior Scientists "Modern achievements of the fundamental sciences for solving actual medical problems", Astrakhan', pp. 272 – 274 (in Russian).

Novykova N.S. 2003. Clinic-immunological characteristics of the liquidators of the consequences of the Chernobyl's NPS in the remote period. Doc. Thesis, Med. Sci., Novosibirsk State Medical Academy, Novosibirsk, 22 p. (in Russian).

Nyagu F.I., Loganovsky K.N. 1998. Nero-psychiatric effects of ionizing irradiation. Kiev, Publ. Chernobylinterinform, 350 p. (in Russian).

Nyagu A.I., Loganovsky K.N., Chuprovsky N.U. et al. 1999. Neuropsychiatric effects in: V.G.Bebeshko, A.N.Kovalenko (Eds). - Book 2. Clinical aspects of the Chernobyl catastrophe. - K.: "MEDECOL" interdisciplinary Scientific and Research Centre BIO-ECOS, Kiev, pp. 154 -186 (In Russian).

Oganesyan N.M., Asryan K.V., Mirydjanyan M.I., Petrosyan Sh.M., Pogosyan A.S., Abramyan A.K. 2002. Estimation of medical consequences of low doses impact on Armenian liquidators of Chernobyl' accident. Abstracts, 3rd Intern. Symp. " Mechanisms of UltraLow Doses action", Moscow, 3 – 6 December, 2002, p. 114 (in Russian).

Okeanov A.E., Cardis E., Antipova S.I. 1996. Health status and follow-up of the liquidators in Belarus. In: "The Radiological Consequences of the Chernobyl Accident". Proc. 1st Intern. Conf., Minsk, Belarus, March 1996, pp. 851 – 859.

Okeanov A.E., Sosnovskaya E.Y., Priatkina O.P. 2004. A national cancer registry to asses trends after the Chernobyl accident. Swiss Med. Weekly, 134, pp. 645 – 649.

Onyschenko N.P., Kokueva O.V., Sof'ina L.I., Hosroeva D.A., Litvynova T.N. Method of the forecasting of the extent of the risk of the development of the chronic pancreatitis among the liquidators of the consequences of the Chernobyl's NPP. Patent 2211449 Russia, MPK {7} G-1N 33/48, G01N 33/50 / - N 2001114065/14; Declaration from 25.05.2001; Publication data 27.08.2003, Bull. N 24.

Petrova I.N. 2003. Clinical meaning of the micro circular irregulations under the morbus hypertonicus among the liquidators of the consequences of the Chernobyl's NPP. Doc. Thesis, Med. Sci., Kuban State Medical Academy, Krasnodar, 22 p. (in Russian).

Pilinskaya M.A. 1999. Cytogenetic effects in the somatic cells of faces like the biomarkers of the influence of ionizing radiation in small dozes of persons who suffered in consequence of the Chernobyl's catastrophe. Intern. Jour. Radiat. Med. № 2, pp. 60 – 66.

Pilinskaya M.A., Dybsky S.A., Dybs'ka O.B., Pedan L.R. 2003. Cytogenetic examination of the participants of the liquidation of the Chernobyl's catastrophe consequences with FISH method. Ukran. Medical Science Academy Reports, vol. 9, № 3, pp. 465 – 475 (in Ukrainian).

Ponomarenko B.M., Bobyl'ova O.O., Prokleena T.L. 2002. Present-day state of children's health born by persons suffered in Chernobyl's catastrophe. Ukranian Herald of social hygiene and organization of public health, № 4, pp. 19 – 21 (in Ukrainian).

Popova O.V., Shmarov D.A., Budnik M.I., Kozhinets G.I. 2002. The study of NMR relaxation of blood plasma under the influence of ecological factors of extra-slight intensity. Abstracts, 3rd Intern. Symp. "Mechanisms of Ultra-Low dose action", Moscow, 3 - 6 December, 2002, Moscow, p. 124 (in Russian).

Potapnev M.P., Kuz'menok O.I., Potapova S.M., Smol'nykova V.V., Mislitsky V.F., Rjeutsky V.A., Vasylevskaya T.A., Vahsukhina L.V. 1998. Functional insufficiency of T-cell immunity among the liquidators in 10 years after the catastrophe on Chernobyl NPP. Belarus National Academy Sci. Reports, vol. 42, № 4, pp. 109 – 113 (in Russian).

Preebylova N.N., Sydorets V.M., Neronov A.F., Ovsyannikov A.G. 2004. The results of the examination of liquidators of the consequences of the Chernobyl's catastrophe. Collect. of Papers, 69th Final Sci. Session Kursk State Medical University and Dept.

Med.-Biol. Sci. Central- Black Soil Sci. Center, Russ. Acad. Med. Sci., Kursk, part 2, pp. 107 – 108 (in Russian).

Problems of the ensuring of the ecological and of the radiation- sanitary safety of the regions suffered from the radiation pollution. 2002. «Russian Ecological Safety». Materials Ineragency Comm. Of Ecological Safety, RF Security Council (September 1995 – April 2002), part 4, Moscow, pp. 178 – 203 (in Russian).

Prokopenko N.A. 2003.Patholgy of the cardiovascular and of the nervous systems as a consequences of the synergism of radiation affection and of the psycho-emotional tension of the persons suffered from the catastrophe on the Chernobyl's NPP. Aging and Longevity Problems, vol. 12, № 2, pp. 213 – 218 (in Russian).

Pryayazhnyuk A.E., Greeschenko V.G., Fedorenko Z.P., Fedorenko V.A., Gulak L.O., Fuzik N.N., Slypenjuk E.M.,Bormasheva I.V. 1999. Epidemic study of the malignant neoplasms among the liquidators of the consequences of the Chernobyl's catastrophe. Intern. Jour, Radiat. Med., № 2, pp. 42 – 50 (in Russian).

Regulation of the Constitutional court of RF about the case of checking of the constitutionality of some states of the 1st asset of the Federal law from the 24th November 1995 "About the carrying in of changes and of additions into the Law of RF "About the social guard of citizens exposed to the influence of the radiation because of the Chernobyl's catastrophe". (http://www.ksrf.ru/doc/postan/p18_97.htm) (in Russian).

Romanenko N.I., Bobrova V.I., Nemchinova T.G., Golovchenko Yu. I. 1995. The features of the influence of small dozes of ionizing radiation on the condition of nervous system. Materials Intern. Conf. 24 – 28 May, 1995, Kiev, Ukraine, Kiev, p. 264 (in Russian).

Romanenkova V. 1998. Russia – Chernobyl – Liquidators – Health. Joint Lent of News, TASS, 24 April, 1998 (rv/lp 241449 APR 98).

Romanova G.D. 2002. Features of the cerebral hemodynamics and of the functional state of brain among liquidators of the consequences of the Chernobyl's catastrophe. Doc. Thesis, Med. Sci., All-Russian Center for Extreme and Radiat. Medicine, Sankt-Peterburg, 17 p. (in Russian).

Rud' L.I., Dubinkyna V.O., Petrova I.N. Kolomijtseva N.E. 2001. State of the blood stream in the supratrochlear artery and of the vegetative regulation under the arterial hypertension among the liquidators of the consequences of the Chernobyl's catastrophe. «New Thech. of Eyes' Micro-Surgery", Materials 12th Sci.- Pract.. Conf., Orenburg, 14 November, 2001, Orenburg, pp. 298 – 299 (in Russian).

Rumyantseva G.M., Chinkina O.V., Levina T.M. 2002. Psychosomatic aspects of the development of the psychical violations of the liquidators of the consequences of the Chernobyl's catastrophe. Psychotherapy and Psychopharmacology, vol. 4, № 1, pp. 1 – 5 (www.consilium-medicum.com/media/psycho/02_01/7.shtml) (in Russian).

Seenyakova O.K., Rjeutsky V.A., Vasylevich LM. 1997. The analysis of some indexes of the state of health of children of the participants of the liquidation of the Chernobyl's catastrophe consequences. "Actual Questions Med. Rehab. Suffers of Chernobyl' Catastrophe", Materials Sci.-Pract. Conf., devoted 10th anniversary Republ. Hospital of Radiat. Medicine (Minsk, 30 June, 1997), Minsk, pp. 44 – 45 (in Russian).

Serduk A.M., Bobyleva O.A. 1998. Chernobyl and the health of the Ukraine population. 2nd Intern. Conf. «Remote med. Consequences of the Chernobyl Catastrophe», Kiev Ukraine, 1 - 6 June, 1998 , Kiev, p. 132 (in Russian).

Shevchenko V.A., Semov A.B., Akaeva E.A. 1995. Cytogenetic effects among the persons suffered from the catastrophe on the Chernobyl's NPP. Radiat. Biology, Radioecology, vol. 35, # 5, pp. 646 – 653 in Russian).

Shykalov V.F., Usatiy A.F., Sevintsev Yu.V., Kruglova G.I., Kozlova L.V. 2002. Analysis of sweep-biological consequences of the Chernobyl's catastrophe for the liquidators – employees of Russian Scientific Center "Kurchatov's Institute". Med. Radiology and Radiat. Safety, vol. 47, № 3, pp. 23 – 33 (in Russian).

Shkrobot C.I., Gara I.I., Salij Z.V., Furdela M.Ya. 2003. Features of the clinical passing of the vegetative disfunction and the state of the mineral thickness of bone tissue among the liquidators of the consequences of the Chernobyl's catastrophe. Scientific herald Ternopol. State Med. Academy, № 2, pp. 80 – 81 (in Ukrainian).

Shubik V.M. 2002. Immunologic changes in the remote period after the influence of small doses of ionizing radiation. Abstarcts, 3rd Intern Symp. "Mechanisms of Ultra-Low Dose action", Moscow, 3 - 6 December, 2002, Moscow, p. 154 (in Russian).

Sokolova A.V. 2000. The diagnostics and the treatment of the vegetative sensor polyneuropathy of the liquidators of the consequences of the Chernobyl's catastrophe. Doc. Thesis, Med. Sci. Perm' State medical Academy, Perm', 37 p. (in Russian).

Soloshenko E.N. 2002. State of immune homeostasis among the sick with spread dermatosis, which were exposed to the radiation during the Chernobyl's catastrophe. Ukranian Jour. Hematology and Transfusiology, № 5, pp. 34 – 35 (in Ukrainian).

Stepanenko I.V. 2003. Dependence of the changes of the immunologic indexes on the blood pH among the liquidators of the consequences of the Chernobyl's catastrophe. Laborat. Diagnost., № 3, pp. 21 – 23 (in Russian).

Strukov E.L. 2003. Endocrine control under the cardiovascular diseases and under some disfunctions of endocrine organs among persons exposed to the influence of the factors of Chernobyl's catastrophe. Doc. Thesis, Med. Sci. All-Russian Center for Extreme and Radiat. Medicine, Sankt-Peterburh, 42 p. (in Russian).

Sushkevich G.N., Tsib A.F., Lyasko L.I. 1995. The level of neuropeptides among the liquidators of the consequences of the Chernobyl's catastrophe. «Actual and Future disturbances of psychical health after nuclear catastrophe in Chernobyl», Materials Intern. Conf., Kiev, 24 - 28 May 1995, Kiev, "Chernobyl Doctors" Assoc., p. 70. (in Russian).

Svirnovsky A.I., Shamanskaya T.V., Bakun A.V. 1998. About the hematological and cytogenetic indexes of the persons suffered from the catastrophe on the Chernobyl's NPP. 2nd Intern Conf. "Remote Medical Consequences of the Chernobyl catastrophe», Kiev, Ukraine, 1 - 6 June 1998, Kiev, p. 360 (in Russian).

Talalaeva G.V. 2002. Change of the biological time of the liquidators of the consequences of the Chernobyl's catastrophe. Herald of Kazakh National Nuclear Centre, # 3, 11 p. (in Russian).

Tataurschykova N.S., Seedorovich I.G., Ardabatskaya T.B., Zelenskaya N.S., Polyushkina N.S. 1996. The analysis of the prevalence of the allergic pathology among the liquidators of the consequences of the Chernobyl's catastrophe. Hematology and Transfusiology, vol. 41, № 6, pp. 18 – 19 (in Russian).

Tlepshukov I.K., Baluda M.V., Tsib A.F. 1998. Change of the haemostatic homeostasis among the liquidators of the consequences of the Chernobyl's catastrophe. Hematology and Transfusiology, vol. 43, № 1, pp. 39 – 41 (in Russian).

Troshina O.V. 2004. The features of the cerebral hemodynamics and of the peripheral neuromotor apparatus in the remote period among the liquidators of the consequences of the Chernobyl's catastrophe. Doc. Thesis, Med. Sci., Institute of Total pathology and pathophysiol., Moscow, 23 p. (in Russian).

Tseloval'nykova N.V., Balashov N.S., Efremov O.V. 2003. Prevalence of the diseases of the respiratory organs among the liquidators of the consequences of the Chernobyl's catastrophe. «Prophylactics as base for modern public health», Materials 38th Interregional Sci.-Pract. Conf. of Physicians, Ul'yanovsk, pp. 133 -135 (in Russian).

Tsib A.V., Ivanov V.K. 2000. Medical consequences of the Chernobyl's catastrophe: forecast and factual data of the National register (www.ibrae.ac.ru/russian/register/register.html) (in Russian).

Tsygan V.N., Dudarenko S.V., Antonov V.M., Rosanov M.Yu., Vasil'eva N.A. 2003. Mechanisms of formations of the psychosomatic disorders under low doses impact. Modern Medicine. Theory and Practice, № 5, pp. 16 – 21 (in Russian).

Tukov A.R., Shafransky I.L., Kleeva N.A. 2002. Comparison of the indexes of peripheral blood and of the doze of external radiation among male liquidators of the consequences of the Chernobyl's catastrophe. Med. Radiol. And Radiat. Safety, vol. 47, № 6, pp. 27 – 32 (in Russian).

Tymonin L. 2005. Letters from the zone. Tol'yatty Publ. «Agny», 199 p. (in Russian)

Vartanyan L.S., Gurevich S.M., Kozachenko A.I., Nagler L.G., Burlakova E.B. 2002. Remote consequences of the influence of small ionizing radiation dozes on the condition of enzymatic antioxidant system of people. Reports, 3rd Intern. Symp. "Mechanisms of the Ultra-Low dose action", Moscow, 3 - 6 December, 2002. Radiat. Biology. Radioecology, vol. 43, № 2, pp. 203 – 205 (in Russian).

Yakushin S.S., Smirnova E.A. 2002. Ecological and medical aspects of the radioactive nuclides pneumopathy. Abstracts, All-Russian Sci-Thech. Conf. Students, Young Scientists and Specialists, Ryazan', pp. 2 – 3 (in Russian).

Zabolotniy D.I., Shildovskaya T.V., Rymar V.V. 2001. Hemodynamic irregulations in the carotid system and in vertebrobasilar one among the victims in the result of the Chernobyl's catastrophe. Jour. Ear, Nose and Throat Illness., № 4, pp. 5 – 13 (in Ukrainian).

Zagradskaya O.V. 2002. Clinic-metabolite aspects of the remote consequences of the impact of the radiation among liquidators suffering from CHD. Doc. Thesis, Med. Sci., Perm' Styate Med. Academy, Perm' 24 p. (in Russian).

Zak K.P., Butenko Z.A., Mikhaylovskaya E.V., 1996. Hematological, immunological and molecular-genetic monitoring of the participants of the liquidation of the catastrophe on the Chernobyl's NPP. Abstracts, «Remote Consequences Changes in Immune and Haemopoet. systems», Sci.-Pract. Conf., Kiev, 7-10 May 1996, Kiev, pp. 12 – 13 (in Ukrainian).

Zubovsky G.A., Malova Yu.V. 2002. The Features of the aging of liquidators organisms. Reports, 7th Intern. Sci.-Pract. Conf. "Aging Patient. Quality of Life", Moscow, 1-3 October 2002, Medical Gerontology, vol. 8, № 8, c. 82 (in Russian).

CHAPTER 10
Did Acute Radiation Syndrome Occur Among the Inhabitants of the 30 km Zone?

Tetsuji Imanaka
Research Reactor Institute, Kyoto University
Kumatori-cho, Osaka 590-0494 Japan, imanaka@rri.kyoto-u.ac.jp

Introduction

According to the Chernobyl Forum report of 2005[1], 28 persons died of acute radiation syndrome (ARS) as a result of the Chernobyl NPP accident. These deaths were all among firemen and NPP staffs who worked at the time of the accident and, it was claimed, there was no ARS among inhabitants who were living around the Chernobyl NPP at the time of the accident. This formal opinion has been repeated many times since the first IAEA conference about the Chernobyl accident in August 1986.[2-4] The average external doses of evacuees from the 30-km zone around the Chernobyl NPP were estimated to be 20 and 30 mSv, respectively, for Ukrainian and Belarusian territories. Maximum individual dose was estimated to be about 400 and 300 mSv for Ukrainian and Belarusian evacuees, respectively. Considering a conventional conception that ARS only appears in cases of radiation exposure over 1,000 mSv, they concluded that no ARS *could* occur among the evacuees from the 30-km zone around the Chernobyl NPP.

On the other hand, after the collapse of USSR, evidence appeared in several publications that ARS was observed among inhabitants around Chernobyl. In 1992 Yaroshinskaya disclosed the secret protocols of the Soviet Communist Party containing information about a number of ARS cases among inhabitants around Chernobyl that were reported to the Operative Group of Politic Bureau of Central Committee of the Communist Party in Moscow.[5,6] Excerpts of the information from the protocols are summarized in Table 1. Although the numbers of deaths and serious illnesses described in the protocols closely correspond to those numbers among firemen and plant staff, ARS cases among children means that there were many additional cases of ARS among inhabitants around the Chernobyl NPP. It is also noteworthy that, according to the protocol, on May 6 two infants were hospitalized at No. 6 Hospital in Moscow - the national center for treatment of ARS in the USSR. Lupandin also reported ARS cases among inhabitants after he investigated in 1992 medical records that were kept at the Central Hospital of the Khoiniki district, in the Gomel region of Belarus.[7,8] He reported 8 cases with clear ARS and 82 cases with syndromes considered to be related to radiation exposure.

In this article the possibility of ARS among inhabitants is discussed again, from the point of radiation exposure of the people living around the Chernobyl NPP at the time of the accident.

Radioactivity release and contamination

The explosion at the Chernobyl-4 reactor took place at 01:23 on April 26, 1986. Witness said that there were several sequential explosions, like fireworks in the night sky. Then the subsequent fire began at the reactor and continued for ten days or more, releasing huge amounts of radioactivity into the surrounding environment. According to the USSR report in 1986,[3] a large amount of radioactivity continued to be released up to May 6th, daily release of which is shown in Fig. 1. It should be noted that the released amount in Fig. 1 is decay-corrected to the activity on May 6. That is, considering decay-out due to the half-life of the short-lived radionuclides, the actual release on the first day was about 6 times greater than shown in figure in Fig. 1. On the second and third day they were 4 - 5 times greater than in Fig. 1. So, we can say that radioactive release in the first three days was far more intensive than the impression given in Fig. 1.

Table 1. Excerpts of description from the secret protocols of the Operative Group of the Politic Bureau of the Central Committee of the Communist Party of the Soviet Union.

Date	<Description about the health state of people>
1986 May 4:	By the situation on May 4, 1,882 people are hospitalized in total. Total number of examined people reached 38,000 persons. Radiation disease of various seriousness appeared with 204 persons, including 64 infants.
May 5:	Total number of hospitalized people reached 2,757 persons, including 569 children. Among them, 914 persons have symptoms of radiation disease. 18 persons are in very serious state and 32 persons are in serious state.
May 6:	By the situation at 9:00 on May 6, the total number of hospitalized reached 3,454 persons. Among them, 2,609 persons are in hospital for treatment, including 471 infants. According to confirmed data, the number of radiation disease is 367 cases, including 19 children. Among them, 34 persons are in serious state. In the 6th Hospital in Moscow, 179 persons are in hospital, including two infants.
May 7:	During the last day, 1,821 persons were additionally hospitalized. At 10:00 May 7, the number of persons in hospital for treatment is 4,301, including 1,351 infants. Among them, diagnosis of radiation disease was established with 520 persons, including staffs of Ministry of Internal Affairs of the USSR. 34 persons are in serious state.
May 8:	During the last day, the number of hospitalized persons increased by 2,245, including 730 children. 1,131 persons left hospital. By the situation at 10:00 May 8, the total of 5,415 persons are in hospital for treatment, including 1,928 children. Diagnosis of radiation disease was confirmed with 315 persons.

remark: The total number of 40 protocols are included in the secret document. See http://www.rri.kyoto-u.ac.jp/NSRG/reports/kr21/kr21pdf/IM-data.pdf.

The first radioactive plume released with the explosion of the reactor is considered to have moved in a westerly direction and passed over several kilometers to the south of Pripyat city where about 50,000 NPP workers and their families were living. On the second day (April 27) the main direction of the radioactive plume changed to the north-west and the north. On the third day (April 28) the main direction was considered to be northerly. Then the direction of the radioactive plume changed to the east (April 29) and the south (April 30 and May 1).[9]

Figure 2 shows Pu contamination in the Ukrainian territory.[10] The most contaminated areas are found to be to the west and to the north of the Chernobyl site. Considering both the trend of daily release in Fig. 1 and the direction changes of the radioactive plume, we can suppose that the most serious contamination in the 30-km zone was formed during the first three days after the accident.

Evacuation from the 30 km zone
Although most people in Pripyat city knew of the accident on the first day (April 26), many of them spent that day as a usual Saturday. It was around noon of the second day (April 27) that the order of evacuation was announced with instructions to bring passports and food

for three days. Evacuation of Pripyat city began at 14:00 on April 27th, about 36 hours after the accident. Three hours later Pripyat city became an empty town. [11]

The decision to evacuate the people in other villages and towns within the 30 km zone was made on the seventh day (May 2nd). Evacuation of rural villages, together with their livestock, began on May 3rd. This was a very difficult task and ended on May 14th. In total 116,000 people were evacuated from the 30 km zone around the Chernobyl NPP; 90,000 from the Ukrainian territory, including 45,000 from Pripyat city, and 26,000 from the Belarusian territory.[11,12]

Fig. 1. Daily release rate to the atmosphere of radioactive materials during the Chernobyl accident, excluding noble gases. [3] The values are decay-corrected to May 6, 1986 and are uncertain by ± 50 %.

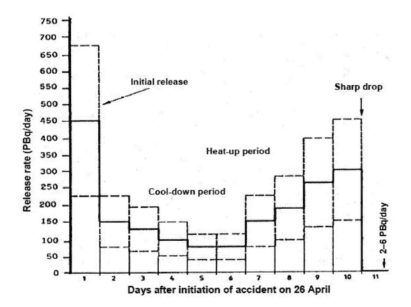

Radiation situation in villages within the 30 km

The detailed radiation situation within the 30 km zone on May 1, 1986 is shown in Fig. 3. This data was published in an EC-Ukraine, Belarus and Russia collaboration report[13] prepared for an international conference held in 1996 in Minsk. The strongest level of 3,306 µGy h[-1] was observed in Krasnoe village about 6 km to the north of the NPP. The second strongest level of 3,045 µGy h[-1] was in the villages of Yanov and Usov. A generally high level of radiation was recorded on the northern side of the 30 km zone. Comparing the contamination pattern of Pu in Fig. 2, it is somewhat strange to note that the high level of radiation was not recorded in villages to the west, over which we can suppose the first radioactive plume passed. To our regret, the radiation situation data, as in Fig. 3, was provided only for May 1st and not provided for other days after the accident.

Fig. 2. $^{238+239+240}$Pu contamination map of the adjacent territory to the Chernobyl NPP (Ukrainian part). [10]

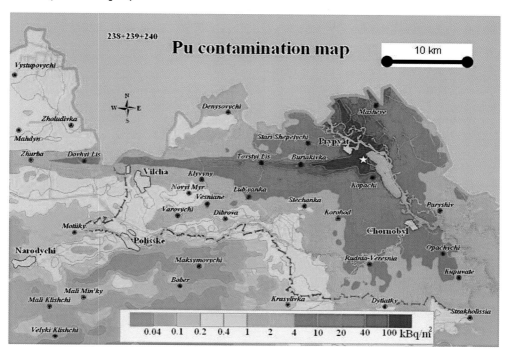

Other interesting data concerning the radiation situation in the 30-km zone was also found in Ref 13. Diamond symbols in Fig. 4 indicate dose-rate change measured in the Khoiniki district by the Belarusian Civil Defense team, the data of which was normalized to unit ^{137}Cs deposition. Two curves are calculated by the present author, assuming two kinds of radionuclide composition deposited on the ground. [14,15] The composition for the solid line is based on the data reported by Izrael et al,[16] while the dotted line was calculated by reducing the ratios for ^{95}Zr and ^{140}Ba so as to fit the measured trend. We can say from Fig. 4 that calculation 2 can reconstruct well the temporal change of the radiation situation during the early period after the Chernobyl accident. We can suppose, therefore, that the dose rate on April 27th was about two times larger than that on May 1st.

Incidentally, when the present author visited Krasnoe village in November 2005, the radiation level there was about 2 μGy h^{-1}, which is 1,500 times less than the value shown in Fig. 3.

External dose estimation by Imanaka

In order to clarify the possibility that the people living within the 30-km zone received a radiation dose that could cause ARS, external doses for evacuees from several villages were estimated based on the above-mentioned data and the following three assumptions:

- Radioactivity deposition occurred at 12:00 on April 27, 1986.
- The dotted line in Fig. 4 can be applied to every village to reconstruct the temporal change of dose rate in air before evacuation.
- A log-normal type of individual dose distribution can be applicable within a village.

Fig. 3. Dose rate in the air at settlements within the 30-km zone around the Chernobyl site on May 1, 1986, unit: µGy/h. [13] Names of settlements were added by the present author.

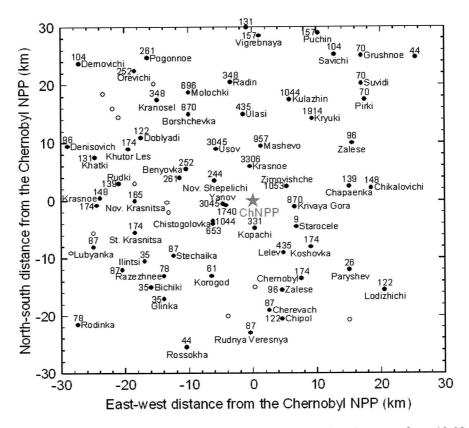

The average dose for each village was calculated by integrating dose rate from 12:00 on April 27 to 12:00 on the day of evacuation. The results of external dose estimation are summarized in Table 2. The detailed procedure used to get values in Table 2 is described in Refs. 14 and 15. The fractions of population whose dose were estimated to be over 500 and 1,000 mSv were calculated in each village based on a log-normal distribution with a geometric SD value of $10^{0.277}$ that was obtained from the data in Ref. 13. The criteria of 500 mSv was chosen as a dose level over which clinically significant depression appears in the blood-forming function.[17] According to Ref. 12, there were 159 inhabitants in Usov village at the time of the accident, which means that 32 and 4 persons could receive more than 500 and 1,000 mSv, respectively. Although the size of the population of Krasnoe village is unclear, judging from the size of the village, its population should be several times larger than Usov. Our results of external dose estimation, therefore, indicate that a substantial number of evacuees received an external dose of more than 500 or 1,000 mSv, which could cause ARS.

Fig. 4. Dose rate change in the Khoiniki district in the first month after the accident normalized per unit deposition of ^{137}Cs. [14]

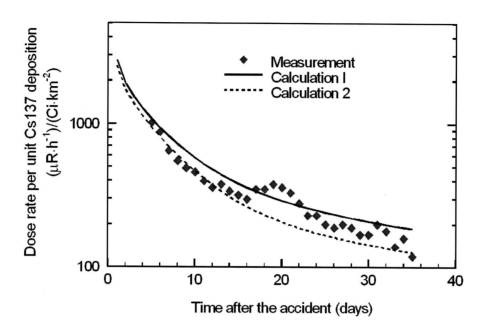

Table 2. Estimates of external dose at several settlements in the 30-km zone until evacuation.

Settlement	Dose rate on May 1, mGy·h^{-1}	Date of evacuation *	Total external dose before evacuation, mSv	Fraction exceeding 500 mSv, %	Fraction exceeding 1,000 mSv, %
Krasnoe	3.306	(May 3)	0.32	24	3.5
Usov	3.045	May 3	0.29	20	2.6
Borshchevka	0.870	(May 5)	0.10	0.5	~0
Chernobyl	0.174	May 5	0.02	~0	~0

Date of evacuation was obtained from Likhtarev et al.[12] were assumed from dates of close-by settlements.

Dose estimation of Chernobyl Forum

The conclusion of the Chernobyl Forum that the average external dose was 20 mSv for evacuees from the Ukrainian part of the 30-km zone is largely based on the work by Likhtarev et al.[12] Using questionnaires regarding the daily behavior of 36,000 evacuees, they reconstructed the individual external dose of about 31,000 evacuees from the 30-km zone, including about 14,000 from Pripyat city. The average external doses of the evacuees were evaluated to be 11.5 and 18.2 mSv for Pripyat city and other villages in the 30-km zone, respectively. The maximum external dose among all evacuees was estimated to be 381 mSv. If we accept these estimates, it is difficult to doubt the occurrence of ARS among inhabitants around the Chernobyl NPP.

The average external dose for evacuees from Usov village was estimated both by Likhtarev et al. (118 mSv) and by the present author (320 mSv). Our estimate is three times larger than Likhtarev et al. Although the reason of the three-fold difference is unclear, several points should be mentioned regarding the work by Likhtarev *et al*. At first, they excluded about 4,000 persons from their dose reconstruction study, the reason being that these people had stayed in highly contaminated areas or visited the Chernobyl station.

Secondly, their average estimate of 18.2 mSv for evacuees from other villages was only 1.5 times larger than that for evacuees from Pripyat city. This ratio seems to be too small considering that the people in Pripyat city evacuated on the second day, while the people in other villages stayed more than a week before evacuation. From the data in the 1986 USSR report [3] the average external doses of 160 and 33 mSv were estimated, respectively, for evacuees from other villages and those from Pripyat city.

Thirdly, although the detailed data of the radiation situation used to estimate external dose in Pripyat city was presented in their paper, no concrete radiation data were shown that were used for other villages. Interesting data is shown in the paper by Muck *et al.*[18] regarding inhalation dose estimation in the 30-km zone, of which Likhtarev is one of co-authors. They plotted temporal changes of dose rate observed during the three weeks following the accident in 49 settlements within the 30-km zone. Considering radioactivity release patterns described in this article so far, temporal changes of dose rate in Muck's paper are very difficult to understand. For example: the maximum dose rate in Yanov village was recorded May 2nd (the seventh day) and that of Krasnoe was May 1st (the sixth day). In all villages dose rates in the first three days (April 26 – 28) were much lower than the later period.

If the dose rate data in Muck's paper were also used in the previous work to estimate external doses, we should consider the possibility that the external dose values in Ref. 12 are underestimated to a large extent, as well as the conclusion of the Chernobyl Forum report.

Conclusion

In order to consider the possibility that ARS occurred among inhabitants around the Chernobyl NPP, external dose was estimated for the evacuees from several highly contaminated villages, using published data of the radiation situation together with several assumptions. Our findings are summarized as follows;

- The average external dose for evacuees from settlements in the 30-km zone did not reach the 500 or 1,000 mSv, criteria for ARS.

- Taking into consideration the distribution of individual doses within a village, substantial numbers (20 to 24 %) of inhabitants in highly contaminated villages could have received external doses exceeding 500 mSv. Some of them could have received more than 1,000 mSv.

- Our results are consistent with the information given in several publications that there were a substantial number of ARS cases among inhabitants around Chernobyl.

- Our estimates are roughly 3 times larger than those made by Likhtarev *et al.*, which is considered to be the basis upon which the conclusion of the Chernobyl Forum report was made: that no ARS occurred among inhabitants.

Although internal dose has not been discussed in this article, one comment should be added. According to the Chernobyl Forum report, the average internal dose of Ukrainian evacuees is concluded to be 10 mSv, excluding 300 mSv of thyroid organ dose. Prohl *et al.* tried to reconstruct the ingestion dose of evacuees from the 30-km zone.[19] They estimated the range of total effective dose by inhalation and ingestion to be 35 – 1,600 mSv for infants and 15 – 440 mSv for adults, more than 50 % of which came from the thyroid dose. These figures clearly indicate that internal dose, which has not been discussed in the present article, contributed significantly to the total dose.

This April, twenty years will have passed since the Chernobyl accident. It must be emphasized, however, that there is still a large contradiction regarding the estimates of radiation dose that the inhabitants living around the Chernobyl NPP received immediately after the accident, as well as contradictions regarding ARS cases among them. Further efforts are needed to resolve these contradictions.

References

1. The Chernobyl Forum; Chernobyl's Legacy: Health, Environmental and Socio-economic Impacts and Recommendations to the Government of Belarus, the Russian Federation and Ukraine. IAEA, September 2005.
2. USSR State Commission on the Utilization of Atomic Energy; The Accident at the Chernobyl Nuclear Power Plant and Its Consequences. August 1986.
3. IAEA; One Decade after Chernobyl: Summing Up the Consequences of the Accident. Proceedings of an International Conference, Vienna, 8-12 April 1996, STI/PUB/1001, IAEA, 1996.
4. UNSCEAR; Source and Effects of Ionizing Radiation, UNSCEAR 2000 Report, Annex J. United Nations, 2000.
5. A. Yaroshinskaya; Chernobyl: Top Secret. Drugie-berega, Moscow, 1992 (in Russian).
6. A. Yaroshinskaya; Impact of Radiation on the Population during the First Weeks and Months after the Chernobyl Accident and Health State of the Population 10 Years Later. In: Imanaka T. ed.; Research Activities about the Radiological Consequences of the Chernobyl NPS Accident and Social Activities to Assist the Sufferers by the Accident Chernobyl. KURRI-KR-21 (http://www.rri.kyoto-u.ac.jp/NSRG/reports/kr21/KURRI-KR-21.htm), 104-107, 1998.
7. V. Lupandin; Invisible Victims. NABAT No. 36, Minsk, October 1992 (in Russian).
8. V. Lupandin; Chernobyl 1996: New Materials concerning Acute Radiation Syndrome around Chernobyl. KURRI-KR-21, 108-113, 1998.
9. Yu. A. Izrael, Chernobyl: Radioactive Contamination in the Environment, Gidrometizdat, 1990 (in Russian).
10. A. Gaydar and O. Nasvit; Analysis of Radioactive Contamination in the Near Zone of Chornobyl NPP. In: Recent Research Activities about the Chernobyl NPP Accident in Belarus, Ukraine and Russia. KURRI-KR-79 (http://www.rri.kyoto-u.ac.jp/NSRG/reports/kr79/KURRI-KR-79.htm) 59-73, 2002.
11. International Advisory Committee; The International Chernobyl Report: Technical Report. IAEA, 1991.
12. I. A. Likhtalev et al.; Retrospective Reconstruction of Individual and Collective External Gamma Doses of Population Evacuated after the Chernobyl Accident. Health Physics, 66(6): 643-652 (1994).
13. I. K. Baliff and V. Stepanenko ed.; Retrospective Dosimetry and Dose Reconstruction. Experimental Collaboration Project No.10, EUR 16540, EC, 1996.
14. T. Imanaka and H. Koide; Dose Assessment for Inhabitants Evacuated from the 30-km Zone Soon after the Chernobyl Accident. KURRI-KR-21, 121-126, 1998.

15. T. Imanaka and H. Koide; Assessment of External Dose to Inhabitants Evacuated from the 30-km Zone soon after the Chernobyl Accident. Radiation Biology Radioecology, 40: 582-588 (2000).
16. Yu.A.Izrael et al.; Radioactive Contamination in the Environment of the Zone around the Chernobyl Atomic Station, Meteorology and Hydrology, 1987 No.2, pp.5-18 (in Russian).
17. 1990 Recommendations of ICRP, ICRP Publication 60, Annals of the ICRP, 21 1991.
18. K. Muck et al.; Reconstruction of the Inhalation Dose in the 30-km Zone after the Chernobyl Accident. Health Physics, 82(2):157-172 (2002).
19. G. Prohl et. al.; Reconstruction of the Ingestion Doses Received by the Population Evacuated from the Settlements in the 30-km Zone around the Chernobyl Reactor. Health Physics, 82(2):173-181 (2002).

CHAPTER 11

Combined Spatial-Temporal Analysis of Malformation Rates in Bavaria After the Chernobyl Accident

Helmut Küchenhoff, Astrid Engelhardt, Alfred Körblein

Malformation rates in the German state of Bavaria, as a whole, did not increase in 1987, the year following the Chernobyl accident. Also an analysis of the monthly data does not show any association between radiation exposure and malformation rates seven months later. But in a detailed analysis on the level of districts, taking the spatial structure into account, we find an association between malformation rates and the calculated caesium concentration in pregnant women. We used a non-parametric estimation of the dose-response relationship, which gives an increasing malformation risk for regions with higher caesium exposure. The results should be interpreted carefully since the analysis has been conducted as an explorative observational study. The results are not in line with the present understanding of the biological effects, including the existence of a threshold dose, for teratogenic effects of low level ionising radiation.

Background

Ionising radiation is an established risk factor for congenital malformations making them an relevant endpoint in the study of possible health effects from the Chernobyl accident. An increased prevalence of congenital malformations at birth was reported for different European countries after the Chernobyl accident. However, existing databases often do not meet quality criteria required for meaningful epidemiologic analysis. The EUROCAT registry only covers approximately 10% of the European population. Also, under-ascertainment of prevalent cases is a systematic problem in many registers.

An overview of the literature on malformation rates following Chernobyl is given in a review article by Hoffmann [1]. In northern Turkey, a significant increase of neural tube defects was reported. A rise in congenital malformations after Chernobyl was found in Belarus. In Croatia, an increase of CNS anomalies was detected in aborted foetuses or in newborns that died within 28 days of delivery. The EUROCAT registry, however, revealed no indication of a systematic increase in the prevalence of Down's syndrome, anencephaly, or spina bifida. Most researchers argued that the radiation exposure from the Chernobyl fallout would be too low to cause any measurable increase in malformation rates.

In Germany, data of the prevalence of malformations at birth were collected *a posteriori* in the State of Bavaria, several years before and after Chernobyl (1984 to 1991). Bavaria was the German Federal state with the highest radiation exposure from Chernobyl. A study, conducted by the German Federal Office of Radiation Protection (Bundesamt für Strahlenschutz, BfS), found no significant difference in malformation rates between the higher exposed southern part of Bavaria and the less contaminated northern part following Chernobyl [2].

In a study on perinatal mortality in Germany following the Chernobyl accident, Körblein and Küchenhoff found a small but statistically significant mortality rise in 1987, the year after the Chernobyl accident. Furthermore, an analysis of the monthly data gave an association between caesium burden and perinatal mortality seven months later [3]. There would have been a similar possible effect on congenital malformations as experimental studies on mice have shown that irradiation of the foetus with 200 R during the period of major organogenesis (day 6 to 13 post conception) resulted in 100% malformed offspring and, to a lesser degree, neonatal deaths [4].

In our study we performed a trend analysis with the same model as in [3]. Then we used ideas of the analysis in [2], in which a comparison of regions was conducted.

Figure 1: Development of caesium concentration in pregnant women following the Chernobyl accident. The light columns are the contributions from milk, the dark columns the additional contributions from beef, pork and cereals.

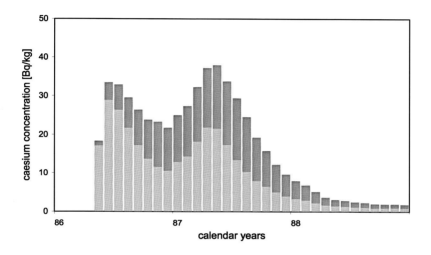

Data

Out of a total of 29,961 newborn with malformations only 7,171 cases were considered appropriate for evaluation; the others were excluded for different reasons by the German Federal Office of Radiation Protection (*Bundesamt für Strahlenschutz, BfS*). Each case was recorded with diagnosis, sex, date of birth and residence of the mother. This data set was provided by *BfS* for evaluation.

Data for the caesium-137 soil contamination on a district level (96 districts) were also obtained from *BfS*. Daily measurements of the caesium concentration in cows milk in Munich from May 1986 until the end of 1988 were provided by the state supported Society for Environment and Health (*Gesellschaft für Umwelt und Gesundheit, GSF*).

Monthly values of the caesium concentration of pregnant women, were calculated on the basis of 4 main food components (milk, beef, pork, cereals) and average consumption rates. They are displayed in Figure 1. There is a first peak in the caesium burden in June/July 1986, and a second in April/May 1987. During the winter 1986/87 cows were fed contaminated grass harvested in summer 1986. Therefore the caesium concentration in pregnant women shows a second increase in the winter season 1986/87.

Figure 2: Estimated effect of the caesium term (CS7) on malformation rates, Bayes' model. The numbers on the y-axis are the logarithms of the odds ratios. The upper and lower curves indicate the 95% confidence limits.

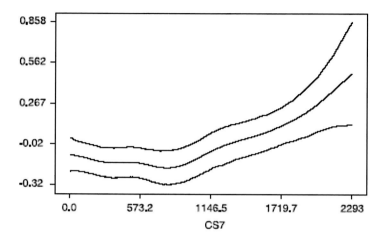

Methods

In a first step, the model used by Körblein and Küchenhoff in [2] is applied to fit the monthly malformation rates in Bavaria:

$$P(Y = 1) = \alpha + \exp(\beta_0 + \beta_1 t) + \beta_2 (cs(t - 7))$$

In a second step, a combined spatial-temporal analysis is conducted. The monthly malformation rates in all 96 Bavarian districts and in 96 months (1984-1991) are analysed as a function of place and time.

As a proxy for the internal caesium exposure in district k at time t, a caesium term cs(k,t) is formed which is the product of the time dependent concentration cs(t) in pregnant women and the caesium soil contamination cs(k) in district k with k = 1, .. , 96. We use a nonparametric logistic regression model of the following form:

$$P(Y(k,t) = 1) = G(\alpha_k + f(cs(k,t - 7) + g(t))$$

Here $G(z) = 1/(1 + \exp(-z))$ is the logistic function.

The probability of a malformation of child i in region k and in month t is denoted by $P(Y(k,t) = 1)$.

The expression f(cs(k,t-7)) is a smooth function which associates the malformation rates with the delayed caesium concentration cs(k,t-7) (time-lag 7 months). This association is estimated by nonparametric methods. Furthermore, the spatial structure is taken into account by assuming a random regional effect which has a correlation dependent on the distance between the regions. The function g(t) is a flexible smooth overall trend component.

For the calculation we use a Bayesian approach by Fahrmeir and Lang [5]. The function f is estimated by a penalised spline approach. The calculations were performed by the program Bayes X [6].

For a sensitivity analysis we fitted other less complicated models, including a parametric logistic model with polynomial trend and polynomial caesium term, and a nonparametric logistic model with a regional effect without assuming a spatial structure. These analyses were performed by the program package SAS. We also conducted the analyses with time-lags of 6 and 8 months between exposure and delivery.

Figure 3: Estimated effect of the caesium term on malformation rates, non-parametric model. The numbers on the y-axis are the logarithms of the odds ratios. The broken lines indicate the 95% confidence limits.

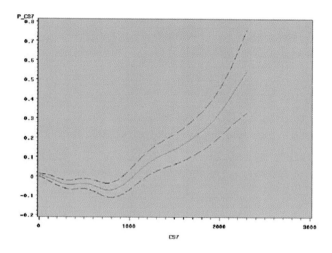

Results
The first model does not give any significant association between caesium exposure and the malformation rates in Bavaria as a whole.

In the second model, the dose response relationship given in Figure 2 is found.

Although formal test procedures have not been performed, the pointwise confidence limits give a clear indication for an association between caesium exposure and malformation risk. The result was confirmed in parametric approaches and in a nonparametric approach without the spatial effect (see Figure 3).

Discussion
To summarise, no association between malformation rates and caesium exposure was found in the first part of the analysis. In a second explorative analysis, using a regression model on an area level, an association between the delayed caesium exposure in pregnant women and malformation rates was found (time-lag 7 months). There is an apparent decrease of risk at low values of the caesium exposure which, however, should not be over-interpreted since the data are also compatible with a practical threshold at low doses. But the decrease at low doses might explain the fact that no caesium effect is found in the data of Bavaria as a whole.

In 1996, the German radiation protection commission (*SSK*) estimated the effective radiation dose from Chernobyl in the highest exposed German areas near the Alps (*Voralpengebiet*) as 0.65 mSv for the first follow-up year. The ICRP 90 recommendations [7] state that the malformation risk is greatest during the period of major organogenesis (3-7 weeks post conception), with an estimated dose threshold of around 100 mSv foetal dose of low-LET radiation, and that the induction of malformations at low doses may therefore be discounted (quoted in the recent WHO report on the Health Effects of the Chernobyl

Accident from July 26, 2005). Since the radiation exposure in Bavaria was 2 orders of magnitude below this threshold dose, no effect on malformations should have been expected.

Our results challenge the concept of a threshold dose of about 100 mSv for the induction of malformations and suggest that the form of the dose response curve at very low doses is non-monotonous. According to Burlakova [8], radiation effects are best characterised by a bimodal dose response curve with a low dose maximum, followed by a plateau region or even a decrease, and a subsequent increase at higher doses. The proposed mechanism involves an increase of DNA repair efficiency by low level radiation exceeding a certain trigger dose.

In conclusion, we believe that the biological effects of very low dose ionising radiation, as yet, are not well understood and deserve further attention.

References

1.   Hoffmann W. Fallout from the Chernobyl nuclear disaster and congenital malformations in Europe. Arch Environ Health. 2001 Nov-Dec;56(6):478-84.

2.   Irl C, Schoetzau A, van Santen F, Grosche B. Birth prevalence of congenital malformations in Bavaria, Germany, after the Chernobyl accident. Eur J Epidemiol. 1995 Dec;11(6):621-5.

3.   Körblein A, Küchenhoff H. Perinatal mortality in Germany following the Chernobyl accident. Radiat Environ Biophys. 1997 Feb;36(1):3-7.

4.   Russell LB, Russell WL. An analysis of the changing radiation response of the developing mouse embryo. J Cell Physiol. 1954 May;43(Suppl. 1):103-49.

5.   Fahrmeir L, Lang S. (2001): Bayesian Inference for Generalized Additive Mixed Models Based on Markov Random Field Priors. Journal of the Royal Statistical Society C, 50: 201-220.

6.   Brezger A, Kneib T, Lang S. (2005): Bayes X: Analysing Bayesian structured additive regression models. Journal of statistical software 14 (11).

7.   ICRP (2003) International Commission on Radiological Protection. Biological effects after prenatal irradiation (embryo and fetus), Publication 90. Annals of the ICRP 23.

8.   Burlakova EB, Goloshchapov AN, Gorbunova NV, Gurevich SM, Zhizhina GP, Kozachenko AI, Konradov AA, Korman DB, Molochkina EM, Nagler LG, Ozerova IB, Skalatskaia SI, Smotriaeva MA, Tarasenko OA, Treshchenkova IuA, Shevchenko VA. [The characteristics of the biological action of low doses of irradiation]. Radiats Biol Radioecol. 1996 Jul-Aug;36(4): 610-31. [Article in Russian].

CHAPTER 12

Radio-Ecological Consequences in Belarus 20 Years After the Chernobyl Catastrophe and the Necessity of Long-Term Radiation Protection for the Population.

Professor V.B. Nesterenko, expert-ecologist A.V. Nesterenko
Institute of Radiation Safety Belrad

1. Analysis of the implementation of short-term radiation protection measures for the Belarusian population

On April 28 and 29, 1986 the Institute of Nuclear Energy of the Academy of Sciences of BSSR (INE) submitted proposals on the implementation of iodine prophylactic measures for the population and on re-settlement of the whole population living within 100-km of the NPP.

In April 1986 our proposals were not accepted and it was not until the beginning of May that the government decided to implement iodine prophylactic measures and to re-settle people from a 30-kilometer zone surrounding the NPP. That May several hundred children were brought to clean regions of Russia.

Following a decision by the government of Belarus, a scientific and technical commission was organized at the beginning of May. It consisted of an academic N.A. Borisevich, the President of the Academy of Sciences, Professor I.N. Nikitchenko, Professor V.B. Nesterenko, Professor Ye.P. Petryayev, Professor S.S. Shushkevich and Professor Ye.F. Konoplya.

On May 3, 1986 I visited the Chernobyl regions of Belarus along with a group of specialists from the radiation safety service of INE. Subsequently, another letter (dated May 7) was sent to the government with proposals to re-settle the population living within the 100-km zone from the NPP, as well as proposals concerning other radiation protection measures.

At the end of May 1986, the first map of Caesium-137 deposition in the Gomel region (fig.1) was drawn up at the Institute. Following the presentation of this data, the population of the southern regions of Belarus (50 to 70-km zone from NPP) was also re-settled, between June 5th and June 10th, 1986.

During the first months after the accident, after making a decision to re-settle inhabitants from affected districts, the local authorities began to implement the following principles; not to allow working resources {sic} to leave their district and to encourage the building of new habitations in these districts. One good example of such a mistake is the decision made by the Mogilyov Regional Committee of the Party and the Regional Executive Committee to build the settlement Maysky in the Cherikov district and to re-settle there the inhabitants from settlements Chudyany and Malinovka of the same district. As a result, the inhabitants of the new settlement raised agricultural products on their former individual plots (3 to 5 km from the new settlement Maysky) with a contamination density over 40 to 80 Ci/km^2

Fig. 1. Caesium 137 deposition in the Gomel region (see text).

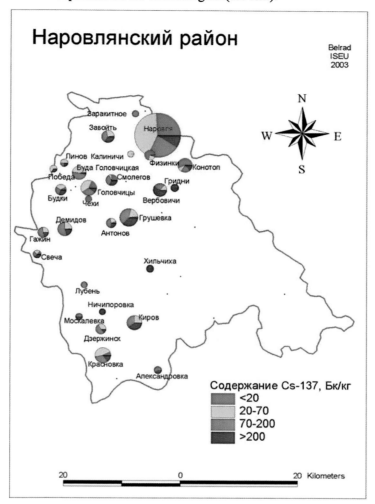

On June 22, 1986 INE submitted the map of Caesium-137 deposition in the Mogilyov region to The Ministry for Public Health Services (MPHS), The Ministry for Agricultural Production (MAP) and to The Mogilyov regional executive committee. As there was no reaction N.A. Borisevich, V.B. Nesterenko and the chairman of the Committee for Hydrometeorology submitted the map to the government and to the Central Committee of the Communist Party of Belarus, proposing that the MPHS investigate fully the possible dangers of living within the 50 settlements of the Mogilyov region. Our proposals were not accepted and no concrete decisions were made, but in September our Institute was visited by the party committee in order to divert the staff of the Institute from activities connected with the Chernobyl subject and to "prove" the effectiveness of their inadequate measures for radiation protection of the population.

In September 1986, by approbation of the authorities of the Academy of Sciences and the government of Belarus, INE submitted a map of radiation deposition in the southern regions of the Republic which included Caesium-137 as well as other radioisotope depositions.

Local foodstuffs were transported from all southern regions of Belarus to the Institute for radiation monitoring. At that time the measurements of radionuclide concentrations in foodstuffs were made at INE, the Institute of Physics of the Academy of Sciences and at The Department for Nuclear Physics of the Byelorussian State University. In the summer of 1986, with the participation of the agrochemical services of the MAP of Belarus, soil samples were taken in all southern regions of Belarus and, in September-October, maps were produced of radiation deposition on agricultural holdings in the southern regions of Belarus (districts and farms).

The population of Belarus that lived in the areas contaminated by Caesium-137 over 37 kBq/m2 consisted of 2,105,200 persons (including over 500 thousand children). To a greater degree a quarter of the territory and one fifth of the population of the republic were affected.

Today the Chernobyl regions of Belarus are characterised by a distorted demographic structure. During the years following the catastrophe 135,000 people were re-settled and not less than 200,000 persons became enforced refugees or left the contaminated district on their own. The first to leave were the young people, intellectuals, skilled specialists and officials. In some affected districts pensioners were to make up about 70% of the population.

Radioactive contamination of the territory caused serious problems in agriculture, especially radiation contamination of agricultural production and local foodstuffs produced on those lands. About 20% (1.6 million hectares) of all agricultural holdings were exposed to ^{137}Cs-contamination over 1 Ci/km^2, largely in traditional agricultural districts. From 1986 to 1990 257,100 hectares of agricultural holdings were excluded from agricultural circulation.

Many errors were made in Belarus during the implementation of activities on contaminated territories, including the lack of effective protective measures for the population. These errors were the result of misguidance by the Joint Government from Moscow and the Byelorussian authorities acting according to the orders of the Governmental Commission on Chernobyl located in Moscow. The Moscow bureaucracy were busy portraying a false picture regarding the safety of living in contaminated regions in order to show the whole world that the dimensions of the damage to Belarus were not significant.

The following are examples of the negligence that ensued; On June 13, 1986 L.N. Kuznetsov, the chairman of the State Committee for Agricultural Production (Gosagroprom) of the USSR , in coordination with P.N. Burgasov, the Chief State Physician of the USSR, accepted "Temporary Recommendations for the Conduct of Agricultural Production in the Byelorussian SSR on Radio-contaminated Territories". This permitted the production of agricultural products even on lands within the 3rd zone where the dose rate was between 5 microroentgens to 20 microroentgens an hour and to distribute them over the whole Republic. The local population living in that area throughout the year accumulated high amounts of radionuclides due to the local dependence on these radio-contaminated foodstuffs.

"Recommendations for the Use of Meat Containing Radioactive Substances 2.0*10-7 to 1.0*10-6 Ci/kg for Production of Cooked Meats for 1986" accepted by the same L.N. Kuznetsov and coordinated with A.I. Zaichenko, the deputy Chief State Physician, were so cruel and cynical. Permissible Caesium-137 concentration levels in sausages were achieved by mixing clean meat with radioactively contaminated meat containing a Caesium-137 concentration between 18,000 and 37,000 Bq/kg.

On June 24, 1986 the same men accepted the "Temporary Recommendations for the Primary Processing of Wool Received from Radio-contaminated Animals". The only factory for the primary processing of wool in the whole republic of Belarus was located in Zhuravichi in the Gomel region; therefore, after simultaneously processing "clean" and "contaminated" wool at that enterprise, all wool eventually became radioactive. It was then used for making clothing, which additionally increased the external dose of the population.

On July 23, 1986 G.A. Romanenko, the deputy Chairman of Gosagroprom of the USSR in coordination with A.I. Zaichenko, the Deputy Chief State Physician, accepted the "Temporary Recommendations for Procedure of Purchase, Acceptance, Storage and Use of Grains and Grassy Meal of the Crop 1986 Gathered on the Territory of the RSFSR, the Ukrainian SSR and Byelorussian SSR Exposed to Radiation Contamination". It was recommended to continue grain production, to continue to feed cattle with the radioactive grain and to continue to use the grain to make alcohol; hence the Caesium-137 contamination of milk and meat, and finally – people.

On the basis of those recommendations Gosagroprom of the BSSR, in association with The Ministry for Grain Production of the BSSR, repeated those instructions in their order No. 3c/21 C dated June 27, 1987. In order to carry out these instructions about 1 million tonnes of radioactive grains were processed and fed at poultry-farms and hog-breeding farms. At the same time two Ministries approved the list of 17 districts of Gomel and Mogilyov regions. This grain should have been exposed to continuous dosimetric control as possible Caesium-137 concentration levels in the grain were 3,700 to 370 Bq/kg.

On July 01, 1987, there was an order by the government of the USSR. This was based on the conclusions of the MPHS of the USSR that it was possible to set a dose limit for the population living on contaminated territories of about 50 rem over 70 years. This was allocated as follows: during the first 10 – 25 years the distribution of the annual doses would be 3; 3; 2.5; 2; 2; 1.5; 1.5; 1; 1; 1 and then 0.5 rem each year up to the 70th year after the accident. In the report on the radiation situation in Mogilyov, Gomel, Bryansk and Kiev regions the Committee for Hydrometeorology of the USSR, MPHS of the USSR, the Council of Ministers of Belarus, the Ukraine and Russia concluded the following – It is possible for the population to continue to live in districts with territorial contamination over 40 Ci/km^2 while using imported foodstuffs (especially milk) through implementing the following precautionary activities:

♦ "decontamination in settlements contaminated over 40 Ci/km^2" (soil layer removal, making firm covers, changing thatches etc.) and the simultaneous ploughing up and sowing of perennial grasses and crops in the fields surrounding these settlements: settlements contaminated over 60 Ci/km^2, to be addressed during 1987, the rest - in 1988);

♦ "Implementation of the whole complex ..... of special soil-conservation activities on all croplands in 1987 and on all pastures in the zone about 40 Ci/km^2 in 1987 and 1988. Separation of plots (in autumn 1987) for the urgent grassing for pastures for de-pasturing cows belonging to the population in settlements contaminated over 40 Ci/km^2 to the cost of croplands (with their exclusion out of land tenure)";

♦ "completing the recommended soil-conservation activities in 1987 on the territories of private farms contaminated under 40 Ci/km^2 and implementation of intensive soil-conservation activities (with the application about 20 tonnes of zeolites per hectare and increased quantities of fertilisers) in settlements contaminated over 40 Ci/km^2";

♦ "implementation of a complex of activities on agricultural holdings contaminated by about 80 Ci/km^2 will lead to a decrease in the contamination of agricultural products by 1987 and 1988, and in a further 3 to 4 years the pollution of all foodstuffs is expected to decrease to the level of the established norms, leading to a cancellation on all restrictions in the use of private agricultural products from private plots (within the first 10 years)".

In March 1988 the chairman of Gosagroprom of the USSR, V. Murakhovsky approved the "Guide for Conducting Agriculture in Conditions of Radioactive Contamination of the Part of the Territory of the RSFSR, the Ukrainian SSR and Byelorussian SSR for 1988 to 1990" submitted by the Inter-departmental commission of scientific experts in radiology and agriculture, concerning peculiarities in the production of agriculture on radio-contaminated territories from 1.5 to 40 Ci/km2.

This guide suggested an application of a 1.5 increased dose of phosphoric and potassium fertilisers on hayfields and pastures annually; the continuation of cattle-breeding; the application of 2 to 3 kg of double super-phosphate and 3 to 4 kg of potassium chloride and sulphate per 100 m2, lime materials and zeolite (200 kg per 100 m2) in order to reduce radionuclide entry in fruits, vegetables and potatoes in vegetable gardens.

The use of local fodder for poultry is not limited - provided they are fattened with clean or only slightly contaminated fodder 1 - 1.5 months before slaughter; there are no restrictions on cattle-breeding and pig-breeding and their fattening, but 1.5 or 2 months before their expected slaughter the cattle should be kept indoors and fattened with clean fodder. As is evident from the guide, the specialists of the Union bodies required a continuation of production on contaminated territories, by any means, in spite of the fact that fattening with contaminated fodder caused the contamination of all kinds of products.

Following the instructions made by the Gomel Agricultural Committee, from 1987 to 1990 the officials of enterprises sent people to the re-settlement zone where grass and grains were sown to be used for feeding cattle.

In September 1989 a group of 92 scientists (incl. 5 Byelorussians) joined that dangerous game connected with the concept of safely living in radio-contaminated territories. They petitioned M.S. Gorbachyov. In that petition they contended that the dose of 35 rem for a life-time, accepted by the National Committee for Radiation Protection (NCRP) of the USSR, was based on the long-term examination of the state of health of the population in Hiroshima and Nagasaki, as well as in the Chelyabinsk district of Russia affected by the accident resulting from the storage of radioactive waste in 1957. In their petition the authors insisted that re-settlement of the inhabitants from the settlement should take place and that they could resume normal living conditions without any restrictions, even if the dose there exceeded the 35 rem life-time dose. Those proposals were approved by IAEA, WHO, NCARE and the UN.

That was in 1989. The scientists both in Belarus and in the Ukraine were not in agreement with the "35 rem during a life-time" concept of the NCRP of the USSR –. The Academy of Sciences of Belarus put forward a new recommendation for living on contaminated territories – 7 rem during a life-time.

MPHS of the USSR (A.Kondrusev) and NCRP of the USSR (L.A. Ilyin) organised a visit of representatives from the International Committee for Radiation Protection and WHO to Belarus. On June 24, 1989, at a meeting in the Academy of Sciences, we were persuaded to accept the "35 rem during a lifetime" concept and, during my report, Dr. Pelleren declared that we should accept 70 and 100 rem during a lifetime because of the lack of financing for providing the population with clean foodstuffs and radiation protection.

The new concept of living on territories contaminated after the Chernobyl accident suggested that the permissible limit of 0.1 rem (1 mSv) a year should be accepted, along with their step-by-step goal: in 1991 – 0.5 rem (5 mSv)/a; in 1993 – 0.3 rem (3 mSv)/a; in 1995 – 0.2 rem (2 mSv)/a; in 1998 – 0.1 rem (1 mSv)/a.

The new zonation was accepted:

<u>Zone of obligatory re-settlement</u>: Caesium-137 – 40 Ci/km^2, Strontium-90 – 3 Ci/km^2, Plutonium – 0.1 Ci/km^2,

<u>Zone of re-settlement</u>: Caesium-137 – 15 to 40 Ci/km^2, Strontium-90 – 2 to 3 Ci/km^2, Plutonium – 0.05-0.1 Ci/km^2, when the annual dose can exceed 5 mSv/a.

But now over 28 thousand persons live in this zone *{sic}*, including 7,000 children. In the Ukraine all people were re-settled from the zone of 15 Ci/km^2.

<u>Zone with the right for re-settlement</u>: Caesium-137 5 to 15 Ci/km^2, Strontium-90 – 0.5 to 2 Ci/km^2, Plutonium – 0.01 to 0.05 Ci/km^2, when the permissible radiation limit for the population exceed 1 mSv/a.

<u>Zone of living with periodical monitoring</u>: Caesium-137 1 to 5 Ci/km^2, when the permissible radiation limit for the population must not exceed 1 mSv/a.

2. Organisation of radiation monitoring of agricultural holdings and foodstuffs

In the summer of 1986 the staff of INE and MAP (2 persons from each organisation) selected soil samples from all kinds of agricultural holdings (5 samples from every 100 hectares) and individual plots. The samples with distinct selective coordinates were brought to INE where they were tested on the γ-spectrometer. By July 15, 1986 over 10,000 samples were brought in. Maps of Caesium-137 deposition on agricultural holdings on farms, districts and regions were made according to the results of their tests. Radiochemical measurements of the samples for Strontium-90 concentration were conducted.

Following the first months after the accident at the Chernobyl NPP the control of foodstuffs from contaminated zones was conducted at three Institutes.

By the middle of July 1986 mobile radiological laboratories were created at Gosagroprom and MPHS of Belarus. Radiometers for controlling foodstuffs were produced at the Institute for Nuclear Energy, the Byelorussian State University and the Institute of Physics and distributed to the radiological services and Institutes of the Ministry for Agricultural Production of the republic. The mobile laboratories, with specialists from MPHS and Gosagroprom, visited all the farms in the Gomel and Mogilyov regions and stated that the majority of plants, fodder and products of animal husbandry were contaminated by Caesium-137 and Strontium-90. Unfortunately, without knowing the actual size of the accident, the authorities in the republican departments and in the regions ordered the production of contaminated products everywhere, based on the Union recommendations. Therefore, radiation control services were organised at all 27 enterprises of the meat industry, 127 dairy enterprises, 114 enterprises of the food industry, 61 enterprises of the Ministry of Grain Production, 56 enterprises of the fruit and vegetable industry and also in 1,200 collective and state farms whose territories had been contaminated by radionuclides. Apart from that, 12 agricultural research institutes, 3 republican, 6 regional, 117 local vet bacteriological laboratories, 188 stations for testing meat, 117 inter-regional laboratories, 6 regional stations for chemisation *{sic}*, 10 pedigree enterprises and 78 poultry-farms were used as places for the implementation of radiation monitoring. The whole radiation monitoring system included 2,122 places.

During these first months, the funds for equipping the laboratories were provided by the departments of the institutes in Minsk, and included the output of several hundreds of radiometers KRVP-3AB from the Lenin Minsk industrial works, which had produced them for equipping atomic submarines. According to our request, a large shipment of radiometric devices SRP-68-01 (about 4,000) was delivered from U-mines and atomic enterprises in Siberia. Radiometers KRP-1, KRVP-3AB, RKP-4SM, SRP-68-01, RUPP, RIS-1 were assembled and produced. In order to equip all centres by the middle of June 1986 the following quantities were necessary: 502 radiometers DP-100, 639 – SRP-68-01 and approximately 300 KRVP devices. By the beginning of June there were only 121 DP-100 and 37 SRP-68-01. By January 01, 1987 there were 189 DP-100, 57 KRVP-3AB, and 799 SRP-68-01.

By October 1986 we prepared 3,077 specialists from the Byelorussian State University and the Academy of Sciences of the Republic for work at the centres for radiation control.

In August 1986 the system of radiation monitoring of foodstuffs at Minsk markets was approved.

In August 1987 the system of radiation monitoring of foodstuffs, agricultural products and the environment of Belarus was approved. Apart from agricultural products, it monitored the contamination of mushrooms, wild berries and herbs.

All maps and findings on radiation contamination levels in foodstuffs were classified as secret. In spring 1989, at the 1st session of the Supreme Soviet of the USSR, due to the initiative of the deputies from Belarus, the Ukraine and A.D. Sakharov, it was decided to remove all details connected with the Chernobyl catastrophe from the secrets list.

Following the removal of the material connected with Chernobyl from the secrets list and following the actions of the state bodies for radiation protection of the population, the inhabitants of Belarus (as well as in the Ukraine and in Russia) began to mistrust all information regarding the size of the accident, the contamination degree of local foodstuffs and the health effects submitted by the state bodies.

Even at that time, the main danger for the population came from the consumption of radio-contaminated foodstuffs. That danger - of constant radionuclide accumulation in the inhabitants of the Chernobyl regions and their internal exposure in small doses - still remains to this day: 17 years after the accident at the Chernobyl NPP.

The Byelorussian writer Ales Adamovich, A.D. Sakharov, the chairman of the Peace Foundation and chess player Anatoly Karpov all suggested that I should organise an Institute for radiation protection for the population of Belarus. First of all, the population should have been informed about the actual radiation situation after the Chernobyl accident, they should have been informed about the radionuclide contamination of foodstuffs and nature gifts {sic}, and they should have been trained in simple radiation protection measures for the inhabitants of the Chernobyl regions.

The Institute of Radiation Safety Belrad (Institute Belrad) was established in 1990. The Institute Belrad suggested to the Supreme Soviet, the Byelorussian Government, and the chairman of the Regional Executive Committees the creation of a network of local centres for the radiation control of foodstuffs for the population (LCRC). Those suggestions were included into the Law of the Republic of Belarus "About the Legal Regime of Territories Contaminated by Radiation as a Result of the Catastrophe at the Chernobyl NPP" with the following text in article 40 "In settlements located in zones of radioactive contamination the State committee of the Republic of Belarus for overcoming the consequences of the catastrophe at the Chernobyl NPP opens LCRC, when it is necessary, under the supervision of local authorities for the implementation of citizens' requirements connected with the testing of foodstuffs and things of general use".

The Institute Belrad developed a β- and γ-dosimeter "Sosna". Its production was organised at the Institute and at the industrial works of Gomel, Borisov and Rechitsa (over 300 thousand devices were produced). At the same time the production of dosimeters RKSB-104 was organised at Minsk industrial works.

Having developed and produced over 1,000 gamma-radiometers RUG-92 the Institute Belrad promoted the equipping of radiological services of MAP, the Ministry of Forestry, the Belarusian Cooperation Union and LCRC with reliable devices with a high sensitivity range for the monitoring of Caesium-137 concentrations in foodstuffs, water and the environment.

In the southern part of Belarus a network of 370 LCRC was organised. The first 30 LCRC were opened due to the financial support of the Peace Foundation of the USSR (A. Karpov) and the Byelorussian Peace Foundation (M. Yegorov). The Chernobyl Committee appointed the Institute Belrad the head organisation for creating and maintaining the LCRC and consulting the population. In such centres, located at schools and at local administration buildings, the population had the opportunity to measure the radionuclide concentration in their foodstuffs and to get objective information about the safety of their use, as well as advice on culinary processing techniques for radionuclide decontamination. The staff of the Institute Belrad, in association with the local authorities, selected candidates from local teachers, doctors, nurses and agronomists. They were educated at our Institute and received certificates. Every month the radiometrists sent reports containing the results of the radiation control of foodstuffs, nature gifts and fodder for the population. The Chernobyl Committee played its part by paying for the production of devices for the LCRC and the radiometrists' wages.

Today, the radiation monitoring data (labelled with the surnames of the owners of the foodstuffs) consists of over 350,000 samples and these data have informed the construction of maps displaying the radiation contamination of milk, berries, mushrooms etc.

Today, the number of LCRC supported by the Chernobyl Committee has been reduced to 40 due to a reduction in financing, another 20 LCRC in Belarus continue to operate due to the financial support of the German Chernobyl initiatives. Contamination density of milk is a paramount risk factor for the health of the people, especially children. According to the data of LCRC in the Gomel region and three districts in the Brest region, about 15% of milk controlled by LCRC had Caesium-137 contamination above the permissible level of 100 Bq/l.

As over 60% of the annual internal dose is received by children due to the use for food of local milk contaminated by Caesium-137, it is considered to be the best indicator for determining the safety of living within contaminated territories. In 2001 the number of settlements producing milk with a Caesium-137 concentration exceeding permissible levels was 326. According to the data of MPHS, milk was contaminated by Caesium-137 over 50 Bq/l (the permissible Caesium-137 concentration level for children's food must not exceed 37 Bq/kg, l) in more than 1,100 villages in Belarus.

3. Permissible levels of radionuclide contamination of foodstuffs and the environment under long-term radioactive contamination of the territory

Based on emergency dose limits of 10 rem during the first year, 5 rem in 1987, 3 rem in 1988, 3 rem in 1989, 0.5 in 1990 (50% of external dose, 50% of internal dose), in 1986, 1988 and 1991, MPHS of the USSR approved temporary permissible levels for Caesium-

137 radionuclide concentration in foodstuffs and drinking water. Table 1 shows temporary permissible levels (VDU-86, VDU-88, VDU-91), republican control levels (RKU-90) and RDU-99) for Caesium-137 concentration in foodstuffs and drinking water.

Table 1. Temporary permissible levels (VDU-86, VDU-88, VDU-91), Republican control levels (RKU-90) and republican permissible levels (RDU-92, RDU-96, RDU-99) for Caesium-137 concentration in foodstuffs and drinking water

Foodstuff	VDU-86	VDU-88	VDU-91	RKU-90	RDU-92	RDU-96	RDU-99
Drinking water	370	18.5	18.5	18.5	18.5	18.5	10
Milk and whole milk products	370	370	370	185	111	111	100
Concentrated milk	7400	1100	1100	370			200
Butter	7400	1100	370	370	185	185	100
Cottage cheese and curd products	3700	370	370	185			50
Meat and meat products	3700	2960	740	592	600	600	500
beef	3700	1850	740	592	600	600	500
mutton pork, poultry and their products	3700	1850	740	592	600	370	180
Vegetable fat	7400	370	185	185	185	185	40
Adipose, margarine	7400	370	185	185	185	185	100
Potatoes, table greens	3700	740	600	592	370	100	80
Bread and bakery	-	370	370	370	185	74	40
Flour, cereals, sugar	-	370	370	370	370	100	60
Vegetables and edible roots	3700	740	600	185	185	100	100
Fruits	3700	740	600	185	185	100	40
Garden berries	3700	740	600	185	185	100	70
Wild berries and preserved food made from them	-	-	1480	185	185	185	185
Tinned vegetables and fruits, juice, honey	-	740	600	185	185	185	185

Fresh mushrooms	-	-	1480	-	370	370	370
Dried mushrooms, dried fruits	-	11100	7400	3700	3700	3700	2500
Other foodstuffs and food additives	-	-	-	592	370	370	370
Herbs, tea	-	-	7400	1850			
Special products for children of all kinds, ready for use	-	1850	185	37			37

The annual food allowance based on VDU-88 permitted internal dose (accepted by MPHS in August 1990) was 0.7 to 0.8 rem/a. RKU-90. This was calculated in such a way that internal dose would be 0.17 rem/a at the possible intake of radionuclides with foodstuffs. In Belarus, current RKU-90 happened to be stricter than those accepted at the beginning of 1991 by MPHS of the USSR's new temporary permissible levels VDU-91.

The principles of local radiation control consist of the obligatory thrice-repeated control: at place of its production, when processing it and, at last, when purchasing integrated products. During the first three-four years, radiation control centres were introduced to all collective farms and state farms and also to all works for keeping, processing and purchasing foodstuffs.

Today, according to the MPHS data, the intake of contaminated foodstuffs has not been reduced, but the statutory permissible exposure levels have been diminished ten times.

During recent years dangerous levels of Strontium-90 contamination in grain, milk and vegetables have been discovered at 28 Belarusian farms.

The Institute for Economics of the Academy of Sciences of Belarus has estimated that, for a 30 year period, the economic damage for Belarus resulting from the Chernobyl catastrophe will be 235 million USA dollars - that is 32 times the annual national budget for the entire republic. Although the state spent from 20% to 6% of the annual budget for Chernobyl programs over the years, this aid to the population of the affected regions was insufficient and did not guarantee the safety of living within the contaminated territories.

The level of income of the inhabitants of these regions is too low for them to buy clean, uncontaminated foodstuffs, thus they are forced to consume food contaminated with Caesium-137. More than 80% to 90% of the annual dose (Caesium-137) is received by the inhabitants because of their dependence on locally produced food.

The effects of long-term small doses of radiation influence negatively on the health of the inhabitants of Belarus, especially children, who live in the Chernobyl regions of the republic.

4. State of health of the population [5]

As a result of the Chernobyl catastrophe the population of Belarus was exposed and is being exposed to the influence of negative factors, primarily radiation. All people within the republic were irradiated by iodine radionuclides during the early period of the accident. About 10,000 people had operations on thyroid cancer, including 1,800 children.

Among the population living or who lived on territories with a Caesium-137 contamination density over 37 kBq/m2 , the scientifically significant increase of the sickness rate by malignant diseases of the respiratory organs, digestive organs and by breast cancer has been confirmed, as well as genetic dysfunctions and congenital malformations.

Other confirmed bodily diseases of the affected population include the scientifically significant increase of the sickness rate by cataract and ischemia, of diseases of the respiratory system, urogenital system, endocrine system, immune system, stomach ulcers, duodenal ulcers and dysfunctions of metabolism.

Tremendous anxiety in this society is caused by the state of its children's health and characterised by the increase of the sickness rate and the decrease in numbers of children classified as healthy (from 85% to 20% in the republic and to 6% in the Gomel region).

5. Principles of radiation protection for the population of Belarus

In Belarus the following laws were accepted;

♦ Law of the Republic of Belarus "About the Social Defence of Citizens Affected by the Accident at the Chernobyl NPP", 1991;

♦ Law of the Republic of Belarus "About the Legal Regime of Territories Contaminated by Radiation as a Result of the Catastrophe at the Chernobyl NPP", 1991;

♦ Law "About Radiation Safety of the Population of Belarus", 1998.

These Laws defined the average annual effective radiation exposure of the population to be of 1 mSv/a. It is written in the addition and the modification of the Law of the Republic of Belarus "About the Social Defence of Citizens Affected by the Accident at the Chernobyl NPP" No 31-1 dated May 17, 2001:

"As an index of the assessment of the territory where the living and working conditions of the population do not require any restrictions, the average annual effective radiation exposure is set that must not exceed 1 mSv above the natural and man-made radiation background.

If the average effective radiation exposure of the population exceeds 1 mSv/a radiation protection activities must take place.

When the average effective radiation exposure of the population is reduced from 1 to 0.1 mSv/a the protective activities must not be cancelled, but their scope and character will be regulated by the Council of Ministers of the Republic of Belarus.

When the average effective radiation exposure of the population is under 0.1 mSv/a above the natural and man-made radiation background the protective activities do not take place and the territory and the population living there is no longer considered under restrictions concerning emergency radiation effects."

It is known that for the dose limit of 1 mSv/a, the Chief Sanitary Inspector of MPHS of the Republic of Belarus approved the dose limits of Caesium-137 and Strontium-90.

Radionuclide concentrations in foodstuffs and drinking water (RDU-99) are based on the annual actual food intake of the inhabitants of the Chernobyl regions. Unfortunately, MPHS did not include the permissible Caesium-137 concentration levels in the basic dose forming foodstuffs corresponding to the dose limit of 0.1 mSv/a. The dose limits are calculated values and can not be measured and are unknown for the population. The set permissible levels of radionuclide concentrations in foodstuffs, equivalent to 0.1 mSv/a, would be a definite guideline for the population for the lower safety limit of contamination of foodstuffs.

In Russia (in 1999) and then in Belarus (in 2000) the following basic National documents in the field of radiation safety and protection of the population were accepted:

♦ Norms of radiation safety and protection of the population.

♦ Basic sanitary rules for radiation prevention.

In Belarus the activities on overcoming the consequences of the accident at the Chernobyl NPP are implemented within the framework of the special state programs financed from the budget. The first program (1990 to 1992) was financed under conditions of the USSR. From 1993 to 1995 and from 1996 to 2000 the republican state programs were implemented. Today the State programs for overcoming of the catastrophe at the Chernobyl NPP for the period 2001-2001 and up to 2010 are in use.

Apart from medical assistance for the population, the most important aspects of the program are the implementation of protective activities on the most contaminated territories, the receipt of exhaustive objective information about the radioactive contamination of the environment, the radiation effects limits for the population, the guaranteeing of agricultural production with radionuclide concentrations which do not exceed the permissible levels, the reduction of radiation effects for the health of people and the substantiation and adjustment of decisions

At one time spectrometers of human radiation (WBC = Whole Body Dosimeters) were placed in all municipal, regional and republican hospitals. The Belarusian government accepted the resolution obliging all heads of enterprises, ministers and departments to make WBC-measurements of all inhabitants of the Chernobyl regions of Belarus. Unfortunately, the low quality of existing WBC services resulted in the annual certification of less than one third of the WBC by the State Committee on Standards. In the Slavgorod district WBC have not been operating for 3 years, in the Bragin district for 2 years.

The WBC-measurements of Caesium-137 accumulation in the inhabitants of these regions characterise the efficiency of the radiation protection activities for the population. The absence of a system for practical WBC-monitoring of Caesium-137 accumulation levels in the children of the Chernobyl regions and for testing their efficiency in the implementation of radiation protection activities is especially pernicious.

Since 1995, the Institute Belrad has begun to create a network of mobile WBC-laboratories for the monitoring of Caesium-137 accumulation levels in 500,000 children in the Chernobyl regions of Belarus. There are now 8 mobile radiological WBC-laboratories, but 15 are necessary.

Since 1996, the Institute Belrad has measured children on WBC to determine their Caesium-137 accumulation. From 1996 to 2003, 190,000 children were measured on WBC. Those measurements demonstrated that only 10 to 15% of children had an internal Caesium-137 accumulation under 10 to 15 Bq/kg. The maximal Caesium-137 concentration levels in children reach 4,000 to 7,200 Bq/kg. The medical investigations made by Doctor of Medicine, Professor Y.B. Burlakova, an academic from the Russian Academy of Sciences A.V. Yablokov (Russia), Professor Yu.I. Bandazhevsky and Professor T.A. Birich (Belarus) demonstrated that pathological dysfunctions of important organs and systems of children can appear at a Caesium-137 accumulation level in the organism of 30 to 50 Bq/kg. Heart muscle is especially sensitive to radiation contamination of the organism.

When having common meals (with radio-contaminated foodstuffs, as a rule) with the family, children receive larger doses because their dose factors are 3 to 5 times higher than those of adults, for this reason the Institute Belrad primarily selects children when performing WBC investigations.

In accordance with the Norms of Radiation Safety (NRS-2000) GN. 2.6.1.8-127-2000, approved by the Ministry for Public Health Services of the Republic of Belarus in 2000 (Article 35) the annual population dose must not exceed basic dose limits (1 mSv/a). These limits are related to the average dose <u>in the critical group of population</u>. In the NRS-2000, in chapter 1, there is the following definition of the critical group within a population; "this is a group of (not less than 10) persons from a population homogeneous in one or several characteristics (sex, age, social and professional conditions, place of residence, food intake) that is exposed by radiation, largely, in the present way of irradiation from the present radiation source".

To provide the statistical significance of the sample ($\sigma = 0.95$), in each settlement the representative group of the population should be measured on WBC. Its number is defined depending on the number of inhabitants, taking into account its social composition and age (not less than 10% of all inhabitants of the present settlement). The critical group is selected according to the results of these WBC measurements. The average value of internal doses of the critical group (but not the average value in the whole settlement) is considered as the annual internal dose for the present settlement and depending on this value a decision on the necessity of radiation protection for the population is taken.

From 1999 to 2004 the Institute Belrad and the Nuclear Research Centre (NRC) "Juelich" (Germany) implemented an international German-Belarusian project "Highly Exposed Children" (over 20 thousand WBC measurements) approved by the Komchernobyl, Gomel Regional Executive Committee, the Ministry for Emergency Situations and The Commission for Problems of the Chernobyl Accident and the House of Representatives of the National Assembly of the Republic of Belarus. In 1996 to 2003, within the framework of the project, the results of WBC measurements of ^{137}Cs in over 145,000 children from 250 villages of Belarus were systematized. Following the results of analysis, in association with the Sakharov Radio-ecological University, maps were constructed of radiation contamination of children from 13 (of 23) districts contaminated as a result of the accident at the Chernobyl NPP. Similar maps were made for the Narovlya, Yelsk, Bragin, Lelchitsy and Chechersk districts. (These maps are reproduced in the annexe to this chapter).

Within the framework of the project with the NRC "Juelich", 20,187 children having the largest ^{137}Cs in their organism were selected for further implementation of WBC-monitoring and implementation of their radiation protection. In each village the critical group (10 persons) having maximum levels of [137]Cs s was selected. This analysis permitted the identification of villages with the most exposed critical groups of children: Svetilovichi of Vetka district (up to 1,536 Bq/kg); Valavsk (up to 744 Bq/kg), Roza-Lyuksemburg (up to 735 Bq/kg), Skorodnoye (up to 682 Bq/kg) of Yelsk district; Lenin (up to 557 Bq/kg) of Zhitkovichi district; Beryosovka (up to 313 Bq/kg), Slobodka (up to 400 Bq/kg) of Kalinkovichi district; Klyapin (up to 667 Bq/kg), Litvinovichi (up to 544 Bq/kg) of Korma district; Dzerzhinsk (up to 254 Bq/kg), Glushkovichi (up to 753 Bq/kg), Grebeni (up to 898 Bq/kg) of Lelchitsy district; Antonov (up to 830 Bq/kg), Verbovichi (up to 1708 Bq/kg), Golovchitsy (up to 743 Bq/kg), Golovchitskya Buda (up to 358 Bq/kg), Grushevka (up to 760 Bq/kg), Demidov (up to 1,090 Bq/kg), Dzerzhinsk (up to 286 Bq/kg), Kirov (up to 1,993 Bq/kg) of Narovlya district; Belyayevka (up to 561 Bq/kg), Polesye (up to 4,240 Bq/kg) of Chechersk district. (See plates r – y)

As a rule, the critical, more vulnerable groups include children from families consisting of many children, troubled families and families with a low income whose food intake contains a great number of radio-contaminated local foodstuffs (milk, forest products, game etc.).

For joint measurements on the assessment of internal dose, age dependant factors, published by the Nuclear Research Centre "Juelich" 05.11.2001 (Germany),were used, although the ICRP constant values, (independent from age K_f) according to the results of the summarizing of the data of the nuclear catastrophe in Japan, 1945, were used in Belarus. It is obvious that the sensitivity of the growing child's organism is, at least, ten times higher than that of adults. Therefore, in the report, experimental WBC value for the level of [137]Cs s, but not calculated values of annual dose exposures of children are used. In figure 2 the dependence of dose factor on age (data of the Nuclear Research Centre "Juelich", Germany) is given and was used for the calculation of internal dose.

The whole scope of WBC measurements of 137Cs s has been performed at the Institute of Radiation Safety Belrad by the staff of the WBC laboratory with 8 mobile gamma-spectrometers ("Skrinner-3M") produced by the Institute for Medical and Ecological Systems (Kiev, Ukraine) in association with the Institute Belrad.

Fig 2; Dependence of dose factor on age based on data from NRC "Juelich" (See text).

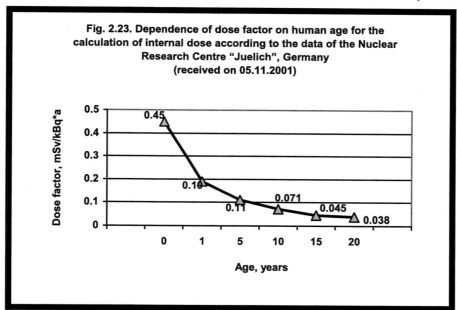

Fig. 2.23. Dependence of dose factor on human age for the calculation of internal dose according to the data of the Nuclear Research Centre "Juelich", Germany (received on 05.11.2001)

The WBC laboratory was certified for independence and technical competence by the National Body for Certification of the Republic of Belarus (Gosstandart) from 2001 to 2004 and from 2004 to 2007. The laboratory is certified annually by the National Body for Certification of the Republic of Belarus. The automated complex "Skrinner-3M" was included in the State Register of Measuring Instruments and was allowed to be used in the Republic of Belarus (All measuring instruments must be verified by the Gosstandard every year and the appropriate certificate is given). All assistants of the laboratory have a higher education, were trained at the firm-manufacturer, received the corresponding certificates and are fully trained in the operation of the complex "Skrinner-3M".

In Table 3 there is minimal detecting activity (MDA) of eight complexes "Skrinner-3M".

Table 3 Minimal detecting activity (MDA) of WBC "SKRINNER-3M" of the Institute Belrad with 30% error and confidence probability for t = 300 seconds and P=0.95 for a 'phantom' man (January 01, 2005)

No.	No. of WBC	MDA, Bq/kg		
		Child (11 kg)	**Teenager** (24 kg)	**Adult** (64 kg)
1	232	13	9	7
2	234	14	9	7
3	240	13	9	6
4	243	13	9	7
5	245	14	8	6
6	246	12	8	6
7	251	12	8	6
8	255	19	12	8

From 2002 to 2004 the WBC laboratory performed triple intercalibrations in association with the laboratory of the Nuclear Research Centre Juelich (Germany) with highly sensitive spectrometers "Canberra" (MDA < 2 Bq/kg). The results of inter-calibrations did not show any deviations from the international standards requirements. The high standards of maintenance on all 8 WBC ensure the scientific accuracy of the results of the measurements made with them.

It is known that, by 1990, at district territorial medical unions within the system of the Ministry for Public Health Services of Belarus and at other Institutes there were 102 WBC. Some of these were designed on the basis of the spectrometer RUP-5 without the necessary biological protections and with poor accuracy MDA, which are of little use for measurements of [137]Cs s in children. Over the last 18 years most WBC owned by the Ministry for Public Health Services have become obsolete and this led to the fact that in 2003, when performing the certification by Gosstandart, 14 out of 29 WBC in the district medical unions did not stand the test and were not certified.

When examining the Checkersk, Stolin, Slavgorod and Bragin districts within the framework of UNDP in 2003, it was ascertained by the CORE group that the WBC in the Bragin district hospital was out of order and had not worked for 2 years. The same situation took place in the Slavgorod district and WBC in Stolin and Checkersk districts should have been substituted urgently.

We can only welcome the decision made by the Swiss Cooperation Office, who placed an order for a new WBC "Skrinner-3M" for the Bragin district, at its own expense, within the framework of the CORE project.

It would be correct to do a full analysis of the technical state of WBC in all districts of Belarus and to plan for their renewal within the next 2 or 3 years.

All measures for radiation and social protection of the population of the Chernobyl regions (re-settlement of inhabitants, implementation of agrochemical measures in agriculture, supplying children at schools and kindergartens with non-contaminated foodstuffs) were insufficient and did not ensure the decrease of ^{137}Cs s in children under 15 to 20 Bq/kg. Therefore, natural biological sorbents for the decontamination of children from radionuclides and heavy metals must be applied.

It must be admitted that information and educational programs for up-grading the radiation knowledge of the population turned out to be insufficient. There is a need within the system for radio-ecological education of the population (especially among children, young people and young parents) at schools, kindergartens, educational establishments and for training in simple methods of radiation protection and means of preventing radionuclide entry into the children's bodies with local foodstuffs. In each district and settlement programs on the introduction of chemical fertilizers to agricultural holdings (including private households), pastures, hayfields and in forests (in places where berries and mushrooms are gathered within a radius of 5 to 10 km from villages) should be performed once every 3 or 4 years for the significant reduction of radionuclide transfer from soil to plants.

In 1992 the Ministry for Public Health Services of the Republic of Belarus published a Catalogue of Dose Burdens of the population for 3,668 settlements within the republic. Unfortunately, when making the catalogue of dose burdens, MPHS made the principle mistake of determining the annual dose burdens in each settlement based on the radioactivity of 10 milk samples and 10 potato samples, this resulted in the inauthenticity of the data and to the depreciation of annual dose burdens in the whole.

It was wrong to use data drawn from legislation aimed at minimising the magnitude of exposure of the population for epidemiological purposes. In determining the average annual dose burden value (collective dose) in order to determine the expected number of patients for purposes of radiation protection, it is important to single out the critical group (10 of the most irradiated persons) and to guarantee such protective measures that the annual dose burdens in the critical group will not exceed 0.3 mSv/a (according to Radiation Safety Norms). In the field of radiation protection there is the principle of the 'critical group' when taking into account protection aimed at the most vulnerable groups and the weakest members of the population (children, pregnant, old people).

In 1998, 2000, 2002, MPHS made attempts to accept the new Catalogue of Dose Burdens of the population of Belarus. The Chernobyl Committee and the House of Representatives of the National Assembly of Belarus organised a commission of experts that carried out direct WBC-measurements of 5,000 inhabitants (mainly children) in 45 villages. Those measurements showed that the calculated dose values (for 10 milk and 10 potatoes samples) of the new Catalogue were underestimated by 6 to 8 times in comparison with the true dose burdens. These results were first reported at the inter-departmental commission and then at the Parliamentary Assembly, the claims were considered as ungrounded and MPHS was forced to remove the new Catalogue of Dose Burdens as inauthentic.

Unfortunately, in summer 2002 MPHS and the National Commission for Radiation Protection (NCRP) introduced new motions to the government of the republic and on August 8, 2002 the Council of Ministers of Belarus approved a new zoning strategy for the territories according to levels of radiation contamination: 146 villages were excluded from the list of objects which are situated within the contaminated territories. These villages were considered to be clean. As a result, the children from these villages were deprived of uncontaminated meals at school canteens and of the annual rehabilitation programs offered at sanatoria in clean regions within the republic, financed by The State Chernobyl Program.

It is very important to develop a system of information and education for the population in order to promote the implementation of scientific and medical recommendations in the field of living safely within the contaminated territories. It is necessary to study not only the sickness rate of the population in the Chernobyl regions, but to implement the necessary radiation protection activities and the treatment of the suffering population.

Listed below are some of the activities that were implemented for minimising internal exposure of the population:

♦ Re-settlement (135 thousand persons were re-settled);

♦ Cultivation of pastures and hayfields, agrochemical protective activities in agriculture (about 0.5 hectares of pastures and hayfields per cow were cultivated only once during a span of 17 years and it is necessary to do this every 3or 4 years);

♦ Mixed fodder with sorbents and boles in cattle-breeding (the production of mixed fodder with sorbents (Prussian blue), calculating 50 kg per cow, was organised at 5 works (200 to 250 kg per cow are needed);

♦ Meals for children at schools and kindergartens due to the Chernobyl Programme, annual rehabilitation in clean districts and abroad;

♦ Organisation of the production of food supplements and their inclusion in the food allowance for decontamination of the organism from radionuclides and heavy metals.

There are some medical works written by scientists from Russia, the Ukraine and Belarus that show that when radiocaesium is 50 to 70 Bq/kg, pathologies of vital systems in children's bodies can appear. The National Report of Belarus 2003 mentions that the main concern "within this society is that of the state of health of the children living within territories with a radiocaesium contamination density of 37 to 555 kBq/m^2 which is characterized by an increase in the sickness rate, a decrease in the number of fully health children and an increase in immune system disorders"

Catastrophic deterioration of the heath of the population has occurred (especially children: In 1985, 85% of children were regarded as fully healthy - in 2001 that figure was reduced to 20%) 20 years after the Chernobyl accident we must conclude that people have not become ill because of stress, radiophobia and over-population, but because of the effects of long-term small radiation doses, because of the continual consumption of radio-contaminated foodstuffs and because of the inability to provide the population of the Chernobyl regions and the whole of Belarus with non-contaminated foodstuffs and other protective measures with the financial resources of one country.

In state structures of the Ukraine, Belarus and Russia the effectiveness of using enterosorbents, for the removal of radionuclides from the bodies of animals in order to treat meat and milk, has already been recognized. In Belarus the production of mixed fodder for cows, containing Prussian Blue, has already been implemented.

It would be beneficial for the population of Belarus to organise the production of pectin food additives for the quick removal of radionuclides from the inhabitants consuming ^{137}Cs contaminated local foodstuffs.

In the Ukraine, food supplements made from apple, beet and citrus polysaccharides, used for the quick removal of radionuclides, are produced at 7 enterprises.

The senior staff at the Centre for Radiation Medicine of the Ministry for Public Health Services of the Ukraine, the Institute for Radiology of the Academy of Sciences of the Ukraine and the staff of the former Institute for Radiation Medicine of the Ministry for Public Health Services of Belarus performed clinical tests on pectin food additives [4, 5], testing their efficiency with the use of WBC, and advised the intake of pectin food additives, for practical radiation protection, not less than 4 times a year (each quarter).

In 1998 the Ministry for Public Health Services of the Ukraine published an information list for the leaders of regional departments for public health services concerning the effectiveness of pectin food supplements "Yablopect" (Dnepropetrovsk, Ukraine) on the inhabitants of radio-contaminated territories and for the recovery of those affected by the Chernobyl accident. We think that this is an important radiation protection measure for children that accelerates the removal and decreases level of radionuclides and heavy metals in children.

The Institute Belrad used the French pectin additive "Medetopect" until 1996 and the Ukrainian pectin preparation "Yablopect" was then used until 2000, in order to decontaminate the bodies of children from radionuclides. From 1998, this work was carried out by the Institute Belrad in association with the international Chernobyl initiatives.

In 1999, within the framework of the joint project with the Austrian Chernobyl initiative, an annual cycle of radiation protection of 1,000 children was implemented in 8 villages. Using WBC controls, these tests revealed a two to four-fold reduction in radiocaesium in children within one year after the use of pectin preparations,

Shown in the annexe is a graph of variations in the average specific activity of 137Cs radionuclides in children from the Sivitsa basic school of Volozhin district of Minsk region.

In Belarus there are sufficient resources of raw materials (apple extraction at 20 canning factories) for the production of pectin food supplements. In April 2000, the Institute Belrad received permission from the Ministry for Public health Services to produce and provide the dried apple vitamin drink "Vitapect" made from apple pectin with additional vitamins B_1, B_2, B_6, B_{12}, C, E, β-carotene, folic acid and microelements K, Zn, Se, Ca. The production of the preparation "Vitapect" was organized at the laboratory of the Institute Belrad at The Charity House.

Following the work of The Chernobyl Committee of Belarus, comparative tests on the efficiency of pectin and vitamin preparations "Medetopect" (France), "Vitapect" (Institute Belrad), Spirulina (Russia) and the vitamin preparation "Vitus-Iod" were carried out at the sanatorium "Belarus". Groups of children (up to 30 persons in each group) took these preparations twice a day over 21 days. WBC-measurements made before and after the intake of these preparations demonstrated a decrease in the Caesium-137 concentration in these children:

- 46 to 49% decrease when taking pectin;
- 31 to 35% decrease when taking Spirulina;
- 23% decrease when taking "Vitus-Iod";
- 18% decrease in the control group (without taking these preparations).

When the effectiveness of the pectin preparation in removing 137Cs from the human organism was called into question, we performed a double blind trial in accordance with international standards. 64 children took part during their recovery at the sanatorium while eating clean foodstuffs.

The association "Children of Chernobyl Belarus" (France) suggested testing the efficacy of the pectin preparation "Vitapect" in comparison with a "Placebo" (based on fruit 'kissel'). The trial took place in June and July, 2001 at the sanatorium "Serebryanye Klyuchi" while consuming uncontaminated food and in accordance with the double blind method. With the cooperation of children and parents, two groups of children were selected. Each group consisted of 32 persons, with one group taking 10 grams of the pectin food additive "Vitapect" daily over 21 days and the second group taking the "Placebo" over the same period. The decrease of ^{137}Cs in children taking the "Vitapect" was 65.6%, those taking the "Placebo" – 13.9%. This statistical difference is scientifically significant (p < 0.01).

From 2000 to 2002 the Institute Belrad, in association with the Chernobyl initiatives from Italy (650 children), Ireland (1,100 children) and England (1,215 children), performed the following projects: children traveling for recuperation were measured on WBC for identification of ^{137}Cs in their bodies; within the recuperation period they took pectin preparations and after their recuperation they were measured on WBC for the second time. A control group was selected that had not received the pectin preparation within the recuperation period abroad. The second WBC measurement (between 26 to 28 days) demonstrated a reduction of the ^{137}Caesium from 60% to 85% in children having taken pectin within the rehabilitation period and 15% to 20% in the control group (those who had not taken the pectin preparation).

In association with the Chernobyl initiatives from Austria, Germany and France, a project was realized that included giving 3 to 4 cycles of "Vitapect", over the course of a year, to children returning from their recuperation. In this study 137Cs in the children were reduced by 2 to 3 times.

A course of radiation protection for 1,400 children in 10 villages of the Narovlya district including the use of "Vitapect" (5 cycles of the "Vitapect" intake and tenfold WBC monitoring of ^{137}Cs in children before and after taking the preparation) was implemented by the Institute Belrad (from 2000 to 2002) together with the association "Children of Chernobyl Belarus" (France), The Mitterrand Foundation and the fund "Children of Chernobyl". This study demonstrated the possibility of reducing annual radiation exposure by 3 to 5 times. The performance of this international project, in coordination with local authorities, incurred little financial costs and is considered to be a good example of the efficiency of such international aid. In the annexe to this chapter are graphs showing variations in the average specific activity of ^{137}Cs radionuclides in children in the villages of Verbovichi and Kirov. These graphs clearly show an increase of 137Cs in children after stopping the "Vitapect" intake.

Among recuperating children in Normandy (France), the French laboratory ACRO tested ^{137}Cs in children's urine before and after taking "Vitapect", and the Institute Belrad performed WBC measurements of ^{137}Cs in children before and after taking "Vitapect". In figure 2.28 in the annexe there is a graph showing the variation of ^{137}Cs in the whole organism before and after taking "Vitapect" over 21 days. The reduction of 137Cs in the organism was 61%.

Over recent years the carers and parents of many hundreds of Belarussian children appeared in Germany (Bremen, Hannover, Bonn, Lüneburg, and Göttingen), Austria (Tirol) and France (Normandy) to obtain pectin preparations. An information certificate is kept for each child that contains the changes of ^{137}Cs in the body (there are three copies– one with the host family, one for the child's family and one kept at the Institute Belrad). Within the last 4 years the Institute Belrad has given the pectin food supplement "Vitapect" to 75 thousand children from the Chernobyl regions of Belarus.

An important property of the pectin preparations should be mentioned: if heavy metals and radionuclides are removed from the organism the positive balance of vitally important microelements is restored. From 2003 to 2004, within the framework of the international project "Highly Exposed Children of Belarus", the efficiency of "Vitapect" vs. the "Placebo" in children was studied over an 8 week period at the sanatoria "Lesnye Dali", "Serebryanye Klyuchi" and "Belorusochka" with the agreement of the children and their parents. The influence of the pectin preparation on blood composition and general recuperation were examined. As well as the three sanatoria, partners for the implementation of the project were the Central Research Laboratory (TsNIL) of the Belarusian Medical

Academy for Postgraduate Education (BELMAPO) of the Ministry for Public health Services of Belarus. Within the radiological section of the project the WBC laboratory of the Institute Belrad performed measurements of ^{137}Cs in children, during their stay, before and after the intake of "Vitapect" and "Placebo". This project consisted of radiological and medical investigations.

Within the radiological section of the project, measurements of 137Cs in children were performed within each recuperation period before and after the intake of "Vitapect" and "Placebo". At the same time radiation control of foodstuffs contributing to the food allowance for children was carried out. In figure 2.30 in the annexe there is an evaluation of the efficiency of "Vitapect" and "Placebo" for the removal of radionuclides from children in the group having taken "Vitapect" and in the control group having taken "Placebo".

The medical part of the project, namely blood sampling, was performed by the medical staff at the sanatorium and the samples were tested by the physicians of TsNIL BELMAPO of the Ministry for Public Health Services of Belarus. A medical checkup of the children was carried out, including functional exercise testing with measurements of blood pressure before and after exercise testing. ECG recordings were made at the beginning and at the end of the recuperation period. Tests for the essential microelements in blood (Zn, Fe, Cu) and for potassium were made at the beginning and at the end of the recuperation period in order to investigate the influence of "Vitapect" on microelementary balance in children.

The application of the pectin preparation "Vitapect" over 12 to 14 days promotes the removal of incorporated ^{137}Cs radionuclides and helps maintain a positive balance of potassium, copper, zinc and iron in children. The reduction of these elements in blood serum has not been observed in tested groups (see annexe).

During their stay at the sanatorium one group of children took the preparation "Vitapect", 5 g two times a day, the second group took the preparation "Placebo" 5 g two times a day. Medical checkups were made twice during the 12 to 14 day interval (at the beginning and at the end of the recuperation period). Presented in the annexe is the diagram showing the responses of the ECG to varying accumulations of Cs-137 in 543 children examined in villages in contaminated districts of Belarus since 2003 and a diagram representing changes in vascular response, with average specific activity of Cs-137.

This work was carried out in accordance with the double blind method of research. The lists of children were decoded after completing all measurements, all medical examinations were conducted in the presence of representatives of the Nuclear Research Centre "Juelich" (Germany).

Before starting the measurements connected with that stage of the project, from October 30 to November 01, 2003, the intercalibration of WBC from the Institute Belrad and the Nuclear Research Centre "Juelich" (Germany) was performed and the certificates of the State Metrological Verification of Spectrometers of the Institute Belrad were examined at the Komchernobyl Children Republican Recuperation Centre "Zhdanovichi".

6. Participation of the non-governmental Institute of Radiation Safety Belrad in overcoming the consequences of the Chernobyl accident in Belarus

During the first years of the Institute Belrad's work, when the Supreme Soviet of the USSR and BSSR made a decision to remove from the secrets list all the material on the Chernobyl accident, the inhabitants of Belarus began to mistrust official government information regarding the levels of contamination in the territory and foodstuffs.

In 1989 the government of Belarus began to acquire staff from the non-governmental Institute of Radiation Safety Belrad (between 1989 to 1990 referred to as the

Scientific and Technical Centre for Radiation Safety "Radiometer") to take part in commissions for the assessment of the radiation situation and the implementation of radiation protection measures for the population. In 1989 and 1990 the Council of Ministries of Belarus (chairman – Professor V.B. Nesterenko), consisting of specialists in radiation protection, agricultural radiology, forestry, medicine and sociology, made thorough investigations of the villages Chudyany (> 147 Ci/km^2), Malinovka and Maysky in the Cherikov district of the Mogilyov region. The proposals of the Commission were accepted by the government and the inhabitants of these villages were re-settled. Six months later the same commission made further investigations of the village of Veprin in the Cherikov district. Following the results of the commission, all children were sent to sanatoria in Russia for several months of rehabilitation, all inhabitants of these village were re-settled, all activity was terminated and the village was land-buried (the streets were contaminated with a Caesium-137 density over 55 Ci/km2).

In 1990 and 1991 the commission investigated living conditions and the possibility of production of foodstuffs containing Caesium-137 within the limits of the RDU-90 (republican permissible levels).Following this, some villages within the Narovlya district went on strike. According to the results of the work of the commission, 65 villages were discovered to have dose burdens in excess of the 35 rem life-time dose. The government made a decision to re-settle 8 villages from the Narovlya district and some of the inhabitants, including their children, from the settlement Narovlya.

In the autumn of 1991 the commission, on behalf of the government of Belarus, made thorough investigations of the villages Ol'myany, Gorodnaya and other villages within the Stolin district. The commission worked closely with local authorities, schools and the entire population. In all villages LCRC for informing the public were organized and additional radiation protection measures for children (uncontaminated meals at schools, year-long double rehabilitation in sanatoria within uncontaminated regions of Belarus etc.) were accepted.

As a result of this work a network of 20 LCRC was organized in the Stolin district, 5 of which were used in implementing the ETOS project.

The next fundamental stage in informing the population, implemented by the Institute Belrad, consisted of developing a program of WBC-radiation monitoring of Caesium-137 accumulation levels in children and discovering ways of lowering these levels.

In 1995 the Institute Belrad began to investigate the feasibility of mobile WBC-laboratories for determining Caesium-137 accumulation levels in children. Due to the help of the Chernobyl initiatives from Germany, Ireland, Norway, the USA and the World Church Council, 8 mobile laboratories were developed using mini-buses, which had been presented to the Institute Belrad by the Irish Chernobyl Children Project and the Vienna City Council.

By the end of the year 2003, over 200,000 WBC-measurements of children were performed at schools and kindergartens in the Chernobyl regions of Belarus. High Caesium-137 accumulation levels in children were discovered; their maximum value reached 4,000 to 7,000 Bq/kg of body weight (BW) per child.

This information was submitted to parents, schools and regional administrations. The results of the WBC-measurements were integrated and published with the permission of the parents and children on the information lists, which were submitted to the Government, Presidents' Administration, the Ministry for Public Health Services (MPHS), all governors, executive committees and all LCRC, to inform them and to discuss protective measures.

On behalf of The Ministry for Emergency Situations (MES) and The Commission of the Chamber of Representatives for the Problems of the Chernobyl Accident, the Institute Belrad was brought in as experts to investigate actual Caesium-137 accumulation levels in the inhabitants of 45 villages, previously declared by the MPHS as 'safe'. The WBC-measurements demonstrated that the actual annual dose burden of the inhabitants of these villages were 6 to 8 times higher than those in the Dose Catalogue submitted by MPHS to the Government of Belarus. In April 2000 the results of the work of the commission were reported at the special session of the Chamber of Representatives for the Problems of the Consequences of the Chernobyl Accident. The proposals of the commission (chairman V.B. Nesterenko) were accepted and MPHS was obliged to recall the draft of the Catalogue of Annual doses for the population of Belarus (2000) from the Government and to send it to the Institute for Radiation Medicine and Endocrinology for revision.

The following day the medical commission came to the Institute Belrad with instructions from the MPHS of Belarus (Minister I.B. Zelenkevich) concerning a prohibition on the Institute Belrad on performing WBC-measurements on the inhabitants. The reason given was that the procedure was, essentially, a medical one and the Institute should have obtained a medical license from MPHS. In a special letter MPHS sent directions to all public health structures ordering the withdrawal of their contracts with the Institute Belrad as it did not possess the necessary medical license.

The Institute Belrad did, however, have a license from the MES of Belarus for the performance of radiation control in the environment and foodstuffs. The decision of the Minster for Public Health Services was appealed in a letter to the President of the Republic of Belarus. On behalf of the Administration of the President, MES carried out an international examination of Belrad's project - "WBC Radiation Monitoring of Radionuclide Accumulation in Children and Their Protection with Pectin Preparations". International experts from Russia, the Ukraine, Belarus and France confirmed that WBC measurements were a physical procedure and the Ministry of Justice of Belarus confirmed that the Institute Belrad, having a license for radiation measurements, did not need a medical license.

MPHS and the Institute Belrad have a data base of WBC measurements of Caesium-137 accumulation levels in children from the Chernobyl regions (maps of radiation contamination of children from 12 districts have been constructed and are shown in the annexe to this chapter). Protective measures should be ensured in such a way that the annual dose burden in the critical group does not exceed 0.3 mSv/a and, in the whole settlement, to remain under 0.1mSv/a, as is stated in the addendum and the modification of the Law of the Republic of Belarus "Regarding the Social Protection of Citizens, affected by the Accident at the Chernobyl NPP" #31-1 dated to May 17, 2001. Provided that adequate financing of the Chernobyl program is in place, such an approach would help to identify the most critical groups, including the most irradiated groups of children from families with many children, and of extended families in each village. In addition, this would help to ensure that the State Chernobyl program could provide the necessary food allowances at schools and annual rehabilitation.

All the results of the WBC-measurements were sent to the local authorities for consultation.

In the Decree of the Council of Ministers of the Republic of Belarus, MPHS was ordered to prepare the Catalogue of Dose Burdens of the population of Belarus. The Republican Scientific and Practical Centre for Radiation Medicine and Human Ecology (RSPC RM&HE) in Gomel prepared the "New Methods of Calculation of the Annual Doses of the Population" which are based on the hypothesis of proportionality between the annual dose burden of the inhabitants of the Chernobyl regions and the contamination density of the territory. The members of the Institute Belrad (Professor V.B. Nesterenko and Professor A.N. Devon) were included in the commission of experts of the National Commission for Radiation Protection for the assessment of the scientific significance of the "Methods of Determining Annual Dose Burdens of the Population of Belarus" proposed by RSPC RM&HE (Gomel). The Institute Belrad submitted to the committee of experts the results of the processed WBC-measurements for 97 villages in the Gomel region, 9 villages in the Mogilyov region and 18 villages in the Brest region. The coefficient K value (relation of the average specific radionuclide accumulation in inhabitants to the contamination density of the territory) varies from 0 to 354 in the Gomel region, from 7 to 95 in the Mogilyov region and from 6 to 85 in the Brest region.

This showed that the main hypothesis of the authors of the "Methods of Determining Annual Dose Burdens of the Population of Belarus" based on the radiation contamination density of soil had no scientific grounding and it was therefore rejected by the National Committee for Radiation Protection.

A new Catalogue of Dose Burdens of the population must be developed, based on the Belgidromet direct measurements of dose rate in each settlement (external exposure dose) and the internal exposure dose of the inhabitants of each settlement should be determined according to the dose burden for the most critical and most irradiated groups of inhabitants (10 to 15 persons). This is received through direct WBC-measurements of the representative group of inhabitants.

At the Gomel RSPC RM&HE the lack of data from direct WBC-measurements of radionuclide accumulation in the inhabitants of several villages of the Gomel region, and especially in the Brest and Mogilyov regions, was conspicious. The Institute Belrad offered the use of its mobile WBC units to the administration of RSPC RM&HE in order to gather this missing information (not less than twice a year in each settlement). The proposal has not yet been accepted.

The measuring abilities of the mobile WBC-laboratories of the Institute Belrad are capable of performing measurements on 55,000 to 60,000 persons a year and ensuring reliable data on the radionuclide accumulation level in the inhabitants of the villages most contaminated by Caesium-137.

Direct WBC-measurements will help to identify the most irradiated groups of the population in the Chernobyl regions of Belarus and to address the issues of radiation protection for the population of the republic.

Since 2000, the Institute Belrad's laboratory installation produces more than 1,500 packs of pectin food supplement "Vitapect" per month. Over the years, due to international help from Chernobyl initiatives, the Institute Belrad has been able to supply "Vitapect" for 75,000 children in the Chernobyl regions of Belarus.

Because of the enlargement of its laboratory facilities in 2005, the Institutes' new "House of Belrad", has been able to produce 5,000 -7,000 packs of "Vitapect" per month. The Irish Chernobyl initiative (the Chernobyl Children's Project) has delivered a press to the Institute for the production of water-soluble tablets containing pectin. A series of foreign Chernobyl initiatives are in place to seek financial help for the expansion of the production of «Vitapect" for radiation treatment for the children of Belarus.

In October 2005, the Institute of Fruit Growing of the National Academy of Sciences of Belarus has offered the Institute Belrad cooperation in the organization and manufacture of pectin food supplements.

Conclusion

1. Implementation of regular WBC measurements on all children in the Chernobyl regions of Belarus is instrumental in identifying nominal critical groups of children with the greatest levels of radiocaesium accumulation in their bodies and opens a pathway to selective priority for radiation protection.

2. It is recommended that the Ministry of Agriculture and Food Production of Belarus should organize, at available canneries within Belarus, the production of pectin food additives from apple-seed meal, and for the Ministry of Public Health and the Ministry of Health of Ukraine to call for the inclusion of pectin food supplements (3-4 times in year) in a regular diet as a means of purification of the bodies of children from radioactive nuclides and heavy metals.

3. There is a need to initiate joint research in order to investigate frequency of disease within the population (especially liquidators and children) relative to Cs-137 content and the influence of radio caesium in their bodies on a host of diseases (heart, kidneys, eyes, endocrine system, etc.).

4. An increase in the number (up to 150-160) of Local Public Centres of Radiation Control of foodstuffs (LCRC) in the contaminated regions is recommended. This would supplement the existing state system of radiation control of food stuffs and could also be an educational base for the population, becoming information centres of radio-ecological education for schoolchildren, their parents and teachers.

5. It is extremely necessary to increase (up to 12-15) the number of mobile WBC laboratories for taking regular measurements of the population and identifying critical groups where dose burdens exceed 1 mSv/year.

6. It is necessary to organize commercial production and regular application of pectin food supplements (made from apples, algae) as an efficient radiation protection treatment for the population of the Chernobyl regions of Belarus.

7. It is necessary to initiate large scale international projects for radiation protection; these include developing the use of pectin food additives and WBC measurement strategies as an aid towards radiation protection for the public.

Experts:

Professor V.B. Nesterenko

Expert-Ecologist A.V. Nesterenko

References

Chernobyl Catastrophe: Causes and Effects. Expert's conclusion of 4 parts (in Russian and English), about 800 pages.

Part 1. Direct Causes of the Accident at the Chernobyl NPP. Dosimetric Monitoring. Protective Activities and Their Efficiency. 1993.

Part 2. Biomedical and Genetic Effects of the Chernobyl Accident. 1993.

Part 3. Effects of the Catastrophe at the Chernobyl NPP for the Republic of the   Belarus. 1992.

Part 4. Effects of the Catastrophe at the Chernobyl NPP for the Ukraine and Russia. 1993.

V.B. Nesterenko. Chernobyl Accident: Radiation Protection of Population, Minsk, 1998, page 172.

I.N. Nikitchenko. Chernobyl. How Does It Happen. Minsk, Committee for Protection of Rights of Industrial Classes, 1999, page 193.

V.F. Ageyets. System of Radioecological Counter-measures in Agrosphere of Belarus. Minsk, 2001, page 249.

Consequences of Chernobyl in Belarus Since 17 Years, National report. Minsk, Committee on Problems of the Consequences at the Chernobyl NPP of the Council of Minister of Belarus, page 52.

Professor, Doctor of Medicine Yu.I. Bandazhevsky. Biomedical Effects of Radiocaesium Incorporated into the Body. Minsk, 2000, page 70.

V.B. Nesterenko. Recommendations on Radiation Protection of Population and Their Efficiency. Minsk, 2001, page 58.

Law of the Republic of Belarus "About the Social Defence of Citizens Affected by the Accident at the Chernobyl NPP", February 22, 1991;

Law of the Republic of Belarus dated 17.05.2001. Addenda and amendments to the Law of the Republic of Belarus "About the Social Defence of Citizens Affected by the Accident at the Chernobyl NPP";

Law of the Republic of Belarus "About the Legal Regime of Territories Contaminated by Radiation as a Result of the Catastrophe at the Chernobyl NPP", November 12, 1991;

I.M. Bagdevich. Guide for Carrying on Agroindustrial Enterprise in Conditions of radiocontaminated Soil of the Republic of Belarus for 1997 to 2000. Minsk, 1997, page 76.

The report of World Bank No 23883-BY "Belarus. The review of consequences of accident on Chernobyl NPP and programs on their overcoming". Minsk, 2002, 64 pages.

The report of the United Nations 2002 "Humanitarian consequences of accident on the Chernobyl NPP. Strategy of rehabilitation". Minsk, Yunipak, 2002, 73 pages.

Professor, Doctor of Medicine M.I. Rudnev. The Opinion About Protective Properties of the Apple Powder Containing Pectins in the experiment of the influence of small doses on the human body. Science centre for Radiation Medicine of Medical Academy of Sciences of the Ukraine, Kiev, 1997.

L.V. Porokhnyak-Ganovskaya. The New Way for Prevention And Rehabilitation of the Inhabitants of the Zone of Radiation Contamination: Pectin Apple Powder And Vitaminized Water-soluble Pills "Yablopect". Medical adviser, No 1, 1998.

N.A. Gres etc. Influence of Pectin Preparations on Dynamics of the Microelement Composition of Blood of Children. Digest of Research Clinical Institute of Radiation Medicine and Endocrinology, Minsk, 1997, pages 108-116.

N.A.Gres, A.N.Arinchik, etc. Peculiarities of the microelement composition of the children's bodies in Belarus. Digest Research Clinical Institute of Radiation Medicine and Endocrinology, Minsk, 1997, pages 26-29.

Joint report IRS Belrad and Research centre in Juelich (Germany) under the international project "Highly irradiated children in Belarus" (4th milestone) "Efficiency of excretion of radioactive nuclides of cesium - 137 from an organism of children by pectin preparation "Vitapect", conservation and stabilization of balance of the vital microelements by it (K, Zn, Fe, Cu). Minsk, Juelich, 2004, 20 pages.

V.B.Nesterenko, A.V.Nesterenko, V.I.Babenko, T.V.Yerkovich, I.V.Babenko. "Reducing the 137-Cs-load in the organism of Chernobyl children with apple-pectin". Swiss Med Wkly, 2004, 134, 24-27.

G.S.Bandazhevskaya, V.B.Nesterenko, V.I.Babenko, I.V.Babenko, T.V.Yerkovich, Yu.I.Bandazhevsky. "Relationship between Cesium (137Cs) load, cardiovascular symptoms, and source of food in "Chernobyl" children – preliminary observations after intake of oral apple pectin." Swiss Med Wkly, 2004, 134:725–729.

V.B.Nesterenko, P.Hill, M.Schläger, H.Dedviichs, R.Lennartz, R.Hille, A.V.Nesterenko, V.I.Babenko. "Evaluation of the Current Radiation Burden of Children Living in Regions Contaminated by the Chernobyl Accident", IRPA-congress at Madrid in Spain, 2004.

ANNEXE TO CHAPTER 12
Figures, maps and diagrams referred to in the text

Belrad contamination data maps

1. Vetka district, village Svetilovichi

The density of contamination of territory by Cs-137 – 22,01 Ci/km^2, dose of external irradiation – 1,0 mSv/a, population – 1042 people (181 schoolboy, in kindergarten – 52 children), quantity of measurements on February 28, 2002 – 152.

Diagram of distribution on intervals of specific activity of radionuclides in organism of inhabitants of village Svetilovichi of Vetka district of Gomel region (28.02.02)

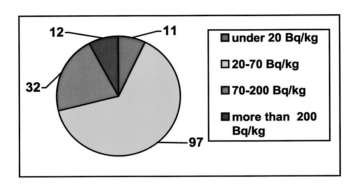

2. Elsk district, village Valavsk

The density of contamination of territory by Cs-137 – 9,44 Ci/km^2, dose of external irradiation – 0,4 mSv/a, population – 819 people (127 schoolboys, in kindergarten – 39 children), quantity of measurements on May 12, 1998 – 213.

Diagram of distribution on intervals of specific activity of radionuclides in organism of inhabitants of village Valavsk of Elsk district of Gomel region (12.05.98)

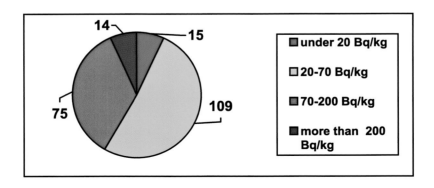

3. Elsk district, village Roza-Luxemburg

The density of contamination of territory by Cs-137 – 10,54 Ci/km^2, dose of external irradiation – 0,5 mSv/a, population – 293 people (72 schoolboys, in kindergarten – 18 children), quantity of measurements on December 5, 2001 – 107.

Diagram of distribution on intervals of specific activity of radionuclides in organism of inhabitants of village Roza Luxemburg of Elsk district of Gomel region (05.12.01)

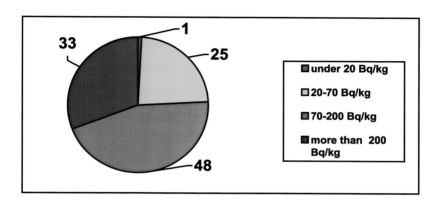

4. Elsk district, village Skorodnoe

The density of contamination of territory by Cs-137 – 6,46 Ci/km^2, dose of external irradiation – 0,3 mSv/a, population – 747 people (200 schoolboys, in kindergarten – 52 children), quantity of measurements on January 11, 2000 – 242.

Diagram of distribution on intervals of specific activity of radionuclides in organism of inhabitants of village Skorodnoe of Elsk district of Gomel region (11.01.00)

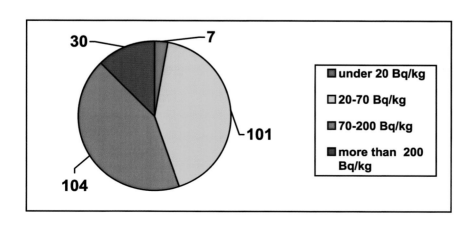

5. Kalinkovichi district, village Beryozovka

The density of contamination of territory by Cs-137 – 4,49 Ci/km^2, dose of external irradiation – 0,2 mSv/a, population – 421 people (120 schoolboys, in kindergarten – 25 children), quantity of measurements on April 19, 2001 – 47.

Diagram of distribution on intervals of specific activity of radionuclides in organism of inhabitants of village Beryozovka of Kalinkovichi district of Gomel region (19.04.01)

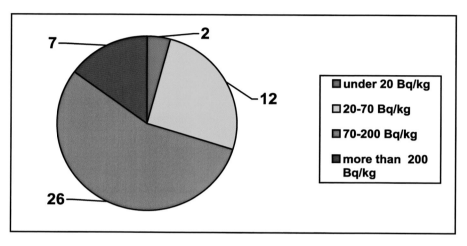

6. Kalinkovichi district, village Slobodka

The density of contamination of territory by Cs-137 – 3,79 Ci/km^2, dose of external irradiation – 0,2 mSv/a, population – 166 people (no school and kindergarten), quantity of measurements on January 27, 2000 – 76.

Diagram of distribution on intervals of specific activity of radionuclides in organism of inhabitants of village Slobodka of Kalinkovichi district of Gomel region (27.01.00)

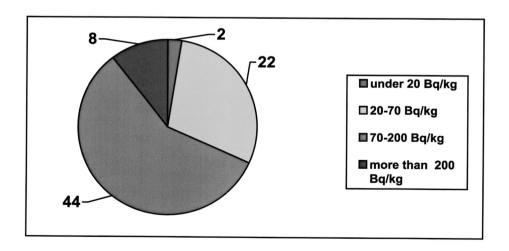

7. Korma district, village Klyapin

The density of contamination of territory by Cs-137 – 6,3 Ci/km^2, dose of external irradiation – 0,3 mSv/a, population – 88 people (49 schoolboys and no kindergarten), quantity of measurements on December 7, 2001 – 50.
Diagram of distribution on intervals of specific activity of radionuclides in organism of inhabitants of village Klyapin of Korma district of Gomel region (07.12.01)

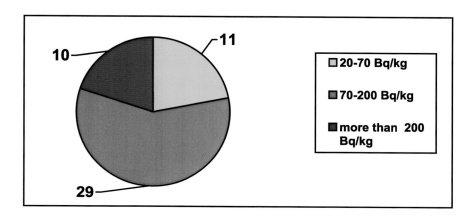

8. Korma district, village Litvinovichi

The density of contamination of territory by Cs-137 – 7,78 Ci/km^2, dose of external irradiation – 0,4 mSv/a, population – 1232 people (280 schoolboys and 44 children in kindergarten), quantity of measurements on September 12, 2001 – 285.
Diagram of distribution on intervals of specific activity of radionuclides in organism of inhabitants of village Litvinovichi of Korma district of Gomel region (12.09.01)

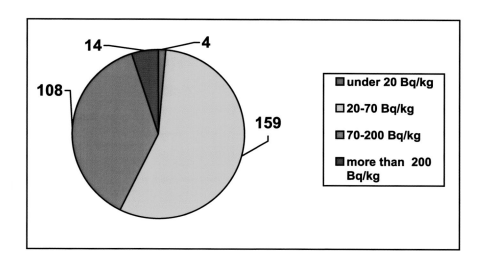

9. Lelchitsy district, village Dzerzhinsk

The density of contamination of territory by Cs-137 – 3,95 Ci/km^2, dose of external irradiation – 0,2 mSv/a, population – 1116 people (190 schoolboys and 100 children in kindergarten), quantity of measurements on January 26, 1999 – 223.

Diagram of distribution on intervals of specific activity of radionuclides in organism of inhabitants of village Dzerzhinsk of Lelchitsy district of Gomel region (26.01.99)

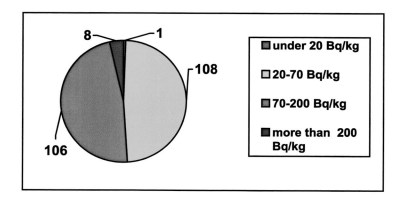

10. Lelchitsy district, village Glushkovichi

The density of contamination of territory by Cs-137 – 2,19 Ci/km^2, dose of external irradiation – 0,1 mSv/a, population – 2349 people (489 schoolboys and 140 children in kindergarten), quantity of measurements on October 24, 2001 – 150.

Diagram of distribution on intervals of specific activity of radionuclides in organism of inhabitants of village Glushkovichi of Lelchitsy district of Gomel region (24.10.01)

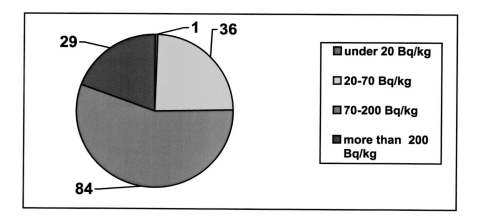

11. Narovlya district, village Verbovichi

The density of contamination of territory by Cs-137 – 19,44 Ci/km², dose of external irradiation – 0,9 mSv/a, population – 396 people (58 schoolboys and 13 children in kindergarten), quantity of measurements on November 27, 2001 – 87.

Diagram of distribution on intervals of specific activity of radionuclides in organism of inhabitants of village Verbovichi of Narovlya district of Gomel region (27.11.01)

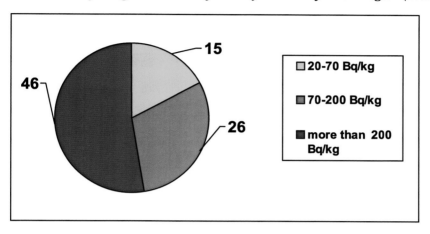

12. Narovlya district, village Golovchitsy

The density of contamination of territory by Cs-137 – 9,44 Ci/km², dose of external irradiation – 0,6 mSv/a, population – 522 people (139 schoolboys and 25 children in kindergarten), quantity of measurements on November 28, 2001 – 160.

Diagram of distribution on intervals of specific activity of radionuclides in organism of inhabitants of village Verbovichi of Narovlya district of Gomel region (28.11.01)

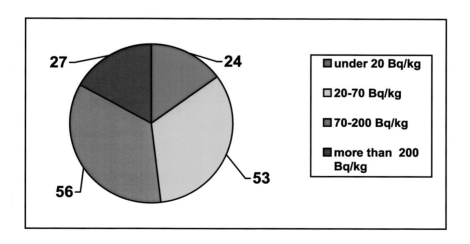

13. Narovlya district, village Demidov

The density of contamination of territory by Cs-137 – 10,38 Ci/km^2, dose of external irradiation – 0,5 mSv/a, population – 311 people (81 schoolboy and 22 children in kindergarten), quantity of measurements on November 13-14, 2001 – 128.

Diagram of distribution on intervals of specific activity of radionuclides in organism of inhabitants of village Demidov of Narovlya district of Gomel region (13-14.11.01)

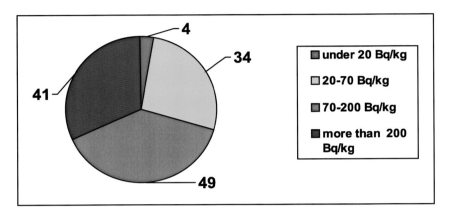

14. Narovlya district, village Kirov

The density of contamination of territory by Cs-137 – 17,39 Ci/km^2, dose of external irradiation – 0,8 mSv/a, population – 424 people (89 schoolboy and 20 children in kindergarten), quantity of measurements on November 28, 2001 – 100.

Diagram of distribution on intervals of specific activity of radionuclides in organism of inhabitants of village Kirov of Narovlya district of Gomel region (28.11.01)

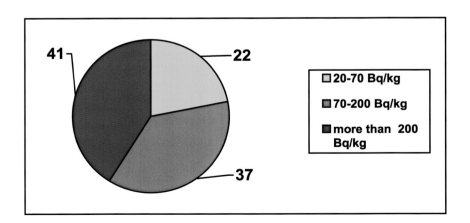

15. Chechersk district, village Belyaevka

The density of contamination of territory by Cs-137 – 6,12 Ci/km^2, dose of external irradiation – 0,3 mSv/a, population – 436 people (70 schoolboy and 15 children in kindergarten), quantity of measurements on February 11, 1999 – 58.

Diagram of distribution on intervals of specific activity of radionuclides in organism of inhabitants of village Belyaevka of Chechersk district of Gomel region (11.02.99)

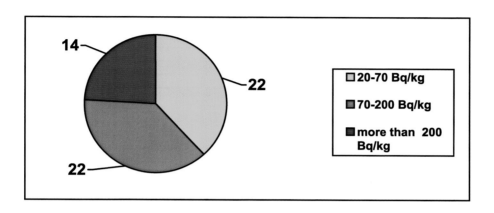

16. Chechersk district, village Polesye

The density of contamination of territory by Cs-137 – 5,7 Ci/km^2, dose of external irradiation – 0,3 mSv/a, population – 566 people (137 schoolboy and 29 children in kindergarten), quantity of measurements on October 24, 2001 – 114.

Diagram of distribution on intervals of specific activity of radionuclides in organism of inhabitants of village Polesye of Chechersk district of Gomel region (24.10.01)

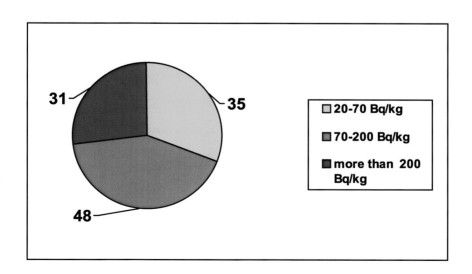

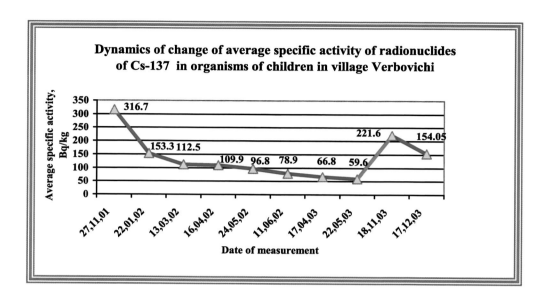

Dynamics of change of average specific activity of radionuclides of Cs-137 in organisms of children in village Verbovichi

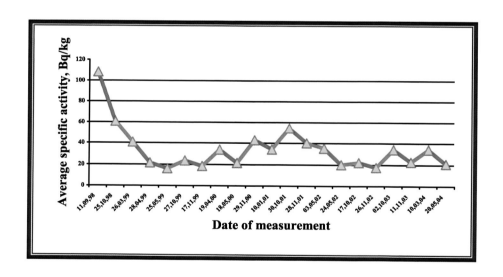

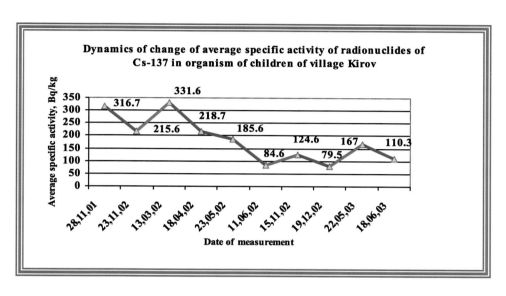

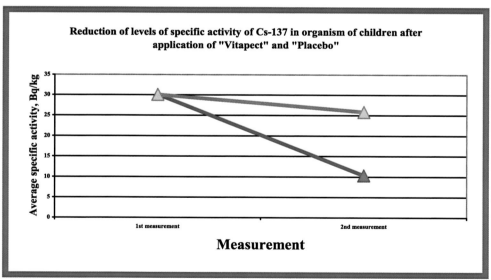

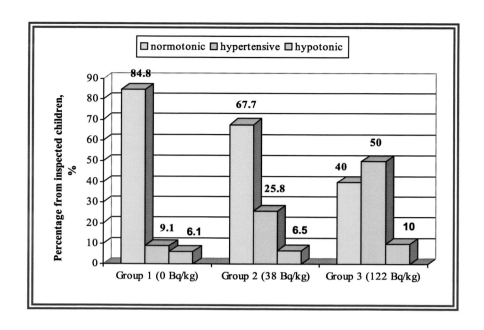

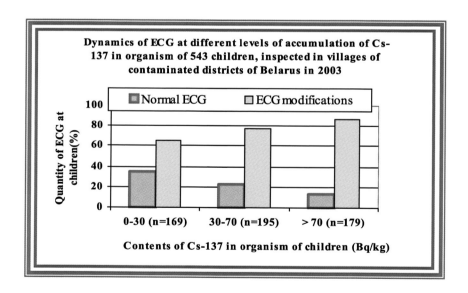

CHANGE-TYPE OF VASCULAR REACTION TO THE LOAD FOR
CHILDREN with AVERAGE SPECIFIC ACTIVITY OF CESIUM - 137

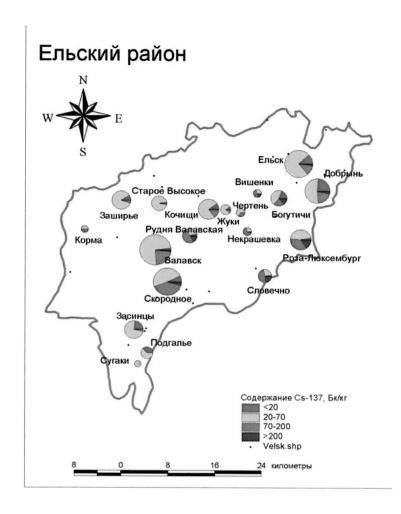

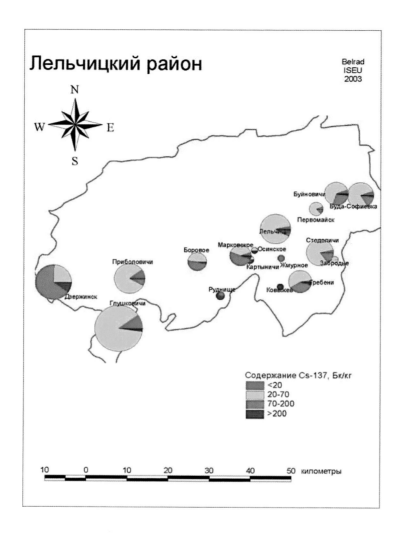

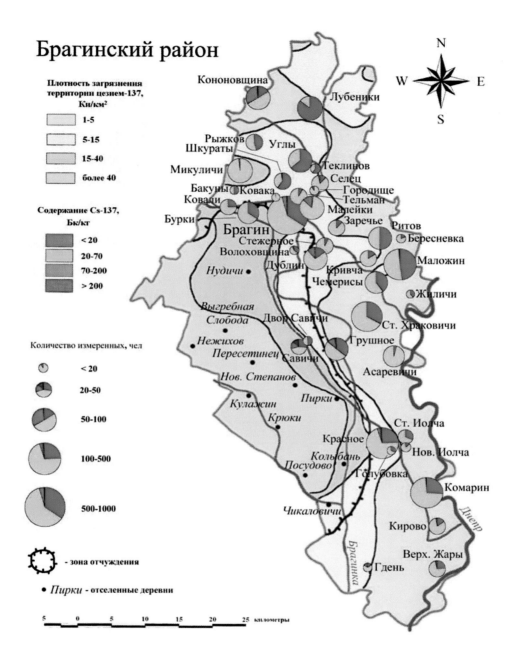

Брагинский район

Плотность загрязнения
территории цезием-137,
Ки/км²

1-5
5-15
15-40
более 40

Содержание Cs-137,
Бк/кг

< 20
20-70
70-200
> 200

Количество измеренных, чел

< 20
20-50
50-100
100-500
500-1000

- зона отчуждения

• *Пирки* - отселенные деревни

5   0   5   10   15   20   25 километры

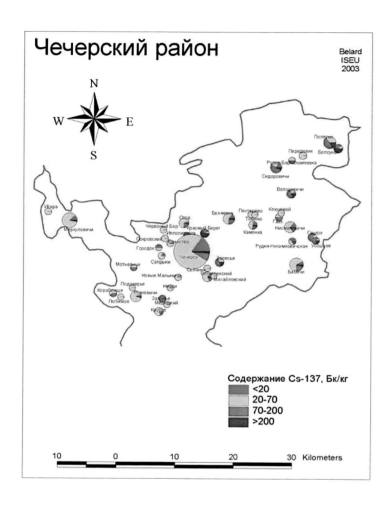

CHAPTER 13

Studies of Pregnancy Outcome Following the Chernobyl Accident

Alfred Korblein

Untere Soeldnersgasse 8, D-90403 Nuremberg, Germany
alfred.koerblein@gmx.de

Perinatal mortality annual data from Germany and Poland show significant increases in 1987, the year following the Chernobyl accident, relative to the long term trend. In an analysis of German monthly perinatal mortality data, peaks of perinatal mortality in the beginning and at the end of 1987 are found. Both peaks are associated with peaks of the caesium burden in pregnant women 7 months earlier (95% CI: 5.5 to 8.5 months). Infant mortality monthly data from Poland exhibit the same pattern. The association with the delayed caesium concentration is highly significant. A combined regression of early neonatal mortality data from Germany and infant mortality data from Poland finds a curvilinear dose response relationship with an estimated power of dose of 2.8 ± 0.8. The perinatal mortality rate in Gomel, the most contaminated region of Belarus, is compared with the rate in the rest of Belarus. The rates do not differ significantly until 1988 but then the rate in Gomel rises and reaches a 30% increased level in the 1990's. This increase can be explained as a late effect of the strontium uptake. In February 1987, a significant 11.4% drop of birth rate is observed in Bavaria which might be explained by an increased rate of spontaneous abortions immediately after Chernobyl when the radiation intensity was highest. The health effects reported here all show a temporal correlation with the radiation exposure from Chernobyl. According to conventional radiobiological knowledge, no teratogenic effects are expected to occur below a threshold dose of about 100 mSv. Even in the most contaminated regions of Germany, however, the extra doses to the foetus were below 1 mSv in the first follow-up year. Therefore the results contradict the widely accepted concept of a threshold dose for radiation damage during foetal development.

1 Background

The observation of pregnancy outcome is a sensitive tool to detect possible adverse health effects on human populations exposed to ionising radiation or other toxic agents. Adverse pregnancy outcomes can be spontaneous abortions, congenital malformations, or perinatal deaths. While perinatal mortality data are available in most countries, there are few registers of congenital malformations, and most registers are incomplete. The European register of congenital malformations EUROCAT covers only about 10% of the European population.

The nuclear accident in Chernobyl on April 26, 1986, was the most severe accident during the civil use of nuclear power. Large parts of Europe were contaminated by the radioactive isotopes of iodine and caesium; strontium and plutonium were also recorded nearer to the Chernobyl site. For radiobiologists and epidemiologists, the Chernobyl accident offered a unique chance of studying the effects of low level ionising radiation on the health of the general population. But the consequences of the Chernobyl disaster are still under debate. The only generally accepted health effect of the Chernobyl accident is a dramatic rise of thyroid cancers in Belarus, Ukraine and parts of the Russian Federation.

Few studies are published in peer reviewed journals on the effects of Chernobyl on perinatal mortality or stillbirth rates. Some of the papers are descriptive reports; precise quantitative results (p-values and confidence intervals) are rather rare. In most cases negative findings are reported, as a rule without mentioning the statistical power of the studies.

In Finland, no significant increase of malformations rates or perinatal deaths was found [1]. In West Germany, an upward deviation from an exponentially falling trend of early neonatal mortality was reported after Chernobyl in Bavaria and Baden-Wuerttemberg, the German Federal States with the highest radiation exposure [2]. But the result depended on the extrapolation model used and was suspected to be an artefact [3]. In Norway, there was an increase of spontaneous abortions for pregnancies conceived during the first 3 months after the Chernobyl accident [4]. Likewise, a 7.2% decrease of birth rate in February 1987 was found in Italy [5]. In Hungary, no increase of congenital abnormalities was observed after the Chernobyl accident, but again, the rate of live births decreased 9 months after May and June 1986 [6]. In the English counties of Cumbria, Clwyd and Gwynedd, where there was heavy rainfall during the passage of the radioactive cloud, no rise of perinatal mortality rates relative to the rest of the country was observed [7]. In a review article about congenital anomalies and other reproductive outcomes the author concludes that there is no consistent evidence of a detrimental physical effect of the Chernobyl accident on congenital anomalies [8]. An increase in unfavourable pregnancy outcome, including a rise of perinatal mortality, was observed in two heavily polluted districts near the Chernobyl reactor site [9]. In Sweden, no change in the rate of spontaneous abortions or congenital malformations occurred in pregnancies exposed at the time of the accident [10]. In Kiev, no pronounced change of perinatal mortality rates was observed after Chernobyl [11].

An increase of developmental abnormalties was found in 5-12 week old human embryos in Belarus after Chernobyl [12]. Also in Belarus, neonates born in heavily contaminated areas were apparently at risk for developing anaemia, congenital malformations and perinatal death [13]. In a review article Goldman states that no significant adverse medical effects other than thyroid cancers in children have been reported in the populations affected by the Chernobyl accident [14]. In Germany, malformation rates [15] and perinatal mortality rates [16] after Chernobyl in the higher polluted southern part of the State of Bavaria were compared to the rates in the less polluted northern part of Bavaria. No significant differences were found. The statistical power of the study, however, was rather weak [17]. A trend analysis of German monthly perinatal mortality rates found peaks that were associated with calculated peaks of caesium concentration in the bodies of pregnant women 7 months earlier [18]. A persistent increase of stillbirth rates after Chernobyl was reported for some eastern European countries outside the former Soviet Union (Sweden, Poland, Hungary, Greece), while no increase was found in central and western European countries [19]. In the State of Bavaria, excess stillbirth rates in 1987 correlated with the level of fallout [20]. A similar study found no correlation between fallout level and stillbirth rate in Finland [21]. In Belarus, perinatal mortality rates in the region of Gomel, which experienced the highest fallout from Chernobyl, were increased in the 1990's relative to the rates in the remainder of the country. The increase was associated with the calculated strontium concentration in pregnant women [22].

Figure 1: Perinatal mortality rates in Germany and long-term trend.

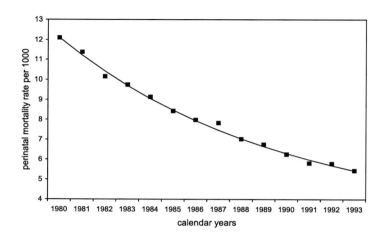

The only official German study of perinatal mortality rates was conducted by the Federal Office of Radiation Protection (Bundesamt für Strahlenschutz) [16]. Data from Bavaria were chosen for this study because Bavaria is the German state that experienced the highest fallout from Chernobyl. The population weighted caesium soil contamination was about 4-times higher in southern Bavaria than in northern Bavaria. A possible radiation effect should have changed the ratio of perinatal mortality in southern to northern Bavaria after Chernobyl. However, the authors did not find any change in this ratio after Chernobyl. This was perceived as evidence that there were no radiation effects from Chernobyl in Germany. It can be shown, however, that the power of the study was too small to detect differences in perinatal mortality rates below 30%. To find small effects, larger populations are needed.

This article is an overview of my studies about the effects of the Chernobyl fallout on pregnancy outcome in Germany, Austria, Poland, Belarus, and Ukraine. To maximise the statistical power of the studies, I included all available German data (80 million population) rather than limiting the investigation to one German region (Bavaria, 11 million population) as in [16]. With a trend analysis, the effects of short time perturbations are investigated on the same population, so any confounding factors can be practically excluded. The calculation of the long-term trend will be more precise as more data before and after 1986, the year of the Chernobyl accident, are available.

2 Data

Perinatal mortality is the number of stillbirths plus early neonatal deaths (0-6 days), divided by the number of live births plus stillbirths. Since in Germany the criteria for stillbirth changed in 1980 and 1994, the study period was restricted to 1980-1993. In 1980, the criterion for stillbirth changed from a body length of 35 cm to a birth weight of 1000 grams, which again was changed to 500 grams in 1994.

German monthly perinatal mortality data, 1980-1993, were obtained from the German Statistical Office (Statistisches Bundesamt). Monthly infant mortality data from Poland, 1985-1990, were supplied by the Institute of Mother and Child in

Figure 2: Perinatal mortality rates in Poland and long-term trend.

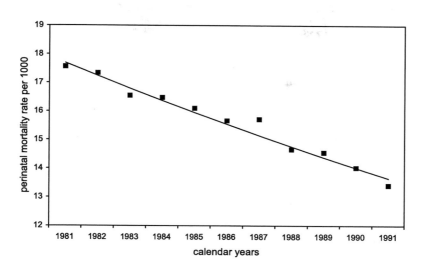

Warsaw. Perinatal mortality data and maternal age distributions from Belarus were provided by the Statistical Department of the Belarus Ministry of Health in Minsk.

The caesium-137 soil contamination ranged from 2-3 kBq/m² in northern Germany to over 50 kBq/m² in southern Germany. Measured caesium concentrations in cows' milk in Munich, Germany, from May 1986 to December 1988, were provided by the state supported Society for Environment and Health (Gesellschaft für Umwelt und Gesundheit, GSF).

3 Annual data

Model

Perinatal mortality data are binary data, therefore the appropriate trend model is a logistic regression model. A flexible form of the time dependency with a linear and a quadratic term is used to describe the undisturbed long-term trend of the data. To test a possible influence of the Chernobyl radiation on perinatal mortality in 1987, a dummy variable d_{87} is used with $d_{87}=0$ in all years except 1987 where $d_{87}=1$. The model has the following form:

$$(3.1) \qquad E(Y(t)) = 1/(1+1/\exp(\beta_1 +\beta_2 \cdot t +\beta_3 \cdot t^2 +\beta_4 \cdot d_{87})).$$

Here $E(Y(t))$ is the estimated perinatal mortality rate, parameter β_1 is the intercept, β_2 and β_3 estimate the linear and quadratic temporal trends, and β_4 estimates the possible increase in 1987. A one-sided t-test is used to determine the significance of the excess perinatal mortality rate in 1987 (hypothesis H_0: $\beta_4 \leq 0$).

Results

Germany: A regression yields a good fit to German perinatal mortality data. The quadratic term in the time dependency is highly significant (p=0.002, F-test). In 1987, there is a significant 4.9% increase relative to the trend of all other years (p=0.0088). The excess rate in 1987 is 0.36 per 1000 births that translates to 317 excess cases (95% CI: 67-578).

The increase of perinatal mortality in 1987 is essentially driven by a 7.4% increase of early neonatal mortality rate (p=0.0035; 95% CI: 2.1% to 13.0%); the 2.9% increase of stillbirth rate is not significant (p=0.100).

Other countries: In the trend analysis of perinatal mortality rates from Poland, 1981-1991, a linear logistic regression model is used. Again, the excess perinatal mortality rate in 1987 is significant (p=0.0074). The excess rate is 0.57 per 1,000 births that translates to 354 excess cases (95% CI: 89-626). About 75% of the excess perinatal deaths in 1987 are early neonatal deaths.

A combined regression for Germany and Poland, with individual trend parameters but a common parameter for the relative increase in 1987, yields a highly significant 4.2% rise (p=0.0003).

The trends of perinatal mortality rates in Germany and in Poland are shown in Figure 1 and Figure 2. The deviations from the trend, in units of standard deviations (standardised residuals), are displayed in Figure 3.

In England and Wales no increase of perinatal mortality rates is observed in 1987 relative to the trend of the data 1981-1992 (see Figure 4). The excess perinatal mortality rates in 1987 obtained from the regressions are listed in the following Table. The increase is greater in Poland than in Germany, and greater in East Germany (former GDR) than in West Germany.

Excess perinatal mortality rates 1987

data set	excess rate	% increase	excess cases	p-value*
Poland	0.572	3.8%	354	0.0074
Germany	0.363	4.9%	317	0.0088
West Germany	0.247	3.5%	159	0.0733
East Germany	0.623	7.2%	141	0.0279

one-sided t-test

4 Monthly data

Model

A model for the trend of monthly data has to allow for seasonal variations of perinatal mortality rates. Two periodic terms with periods of 12 and 6 months are therefore added to model 3.1. Four additional parameters are needed, two for the amplitudes ($\beta4$, $\beta6$) and two for the phase shifts ($\beta5$, $\beta7$):

(4.1) $E(Y(t)) = 1/(1+1/\exp(\beta1 +\beta2{\cdot}t +\beta3{\cdot}t^2 +\beta4{\cdot}\cos(2\pi{\cdot}(t-\beta5)) +\beta6{\cdot}\cos(2\pi{\cdot}(2t-\beta7))))$

The caesium concentration in cows' milk is used as a proxy for the total internal radiation exposure from caesium, essentially because data of caesium contamination in cows' milk were available. The caesium concentration in cow milk was measured nearly every day at the Munich based GSF-Institute, from the beginning of May 1986 until the end of 1988. In the first year following Chernobyl, the internal exposure of the German population exceeded the external caesium exposure [23]. Milk produced in higher contaminated southern Bavaria was distributed and consumed throughout West Germany.

Figure 3: Deviations between observed and expected rates in units of standard deviations (standardised residuals) in Germany (black squares) and Poland (white squares). The broken lines show the range of 2 standard deviations (2σ-range).

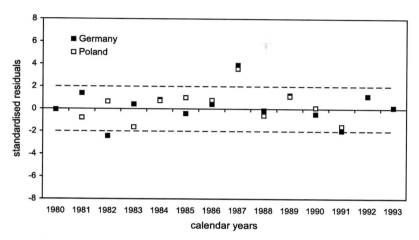

The calculation of the caesium concentration in pregnant women is based on the assumption of a constant daily milk consumption and a somewhat shortened biological half-life of caesium of 70 days during pregnancy. The caesium burden increases with caesium uptake and decreases with caesium excretion. Figure 5 shows the measured caesium concentration in milk (dots) and the calculated caesium burden in pregnant women (solid line).

The regression model with the caesium term has the following form:

(4.2) $E(Y(t)) = 1/(1+1/\exp(\beta1 + \beta2 \cdot t + \beta3 \cdot t^2 + \beta4 \cdot \cos(2\pi \cdot (t-\beta5)) + \beta6 \cdot \cos(2\pi \cdot (2t-\beta7)) + \beta8 \cdot (cs(t-\beta10))^{\wedge}\beta9))$

Parameter $\beta8$ estimates the size of the caesium effect, $\beta10$ is the time-lag between caesium concentration $cs(t)$ and perinatal mortality, and $\beta9$ is the power of dose which allows for a curvilinear dose-response relationship.

To test the significance of the caesium term, the weighted sum of squares resulting from a regression with the full model (4.2) is compared with the sum of squares obtained from a regression without the caesium term (model 4.1). The F-test is applied where the F-value is defined by

$F = ((\chi^2 0 - \chi^2 1)/(df0 - df1))/(\chi^2 1/df1)$.

Here $\chi^2 0$ and $\chi^2 1$ denote the weighted sum of squares under the null hypothesis and under the full model, respectively; df0 and df1 are the corresponding degrees of freedom. Here df0-df1 equals the number of parameters to be tested. The expression $\chi^2 1/df1$ in the denominator is the so called overdispersion factor. For a detailed description of the methods see [18].

Perinatal mortality in Germany

German monthly perinatal mortality rates and the regression line are displayed in Figure 6. Figure 7 shows the deviations of the observed rates from the calculated undisturbed trend (standardised residuals) and the three-month moving average.

Figure 4: Perinatal mortality rates in England and Wales and long-term trend.

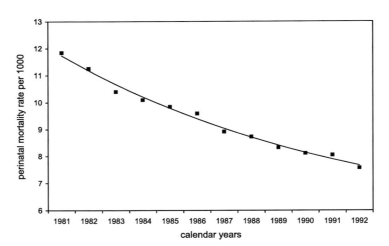

Figure 5: Caesium concentration in cows' milk (short dashes) and in pregnant women (solid line), calculated with a constant daily consumption of cow milk and a biological half life of 70 days.

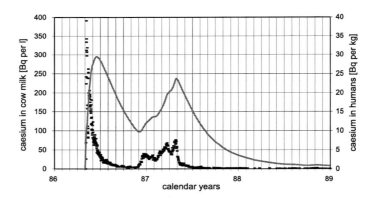

There are significant peaks of perinatal mortality in the beginning and at the end of 1987.

A regression without the caesium term (model 4.1) yields a weighted sum of squares of 221.6 (df=161). Regressions with model 4.2 are then performed with different time-lags β10. The best fit with a sum of squares of 204.5 (df=158) is obtained for a time-lag of 7 months. The corresponding F-test is significant (p=0.0053).

In Figure 8, the profile sum of squares is plotted as a function of the time-lag. The broken line gives an estimate of the 95% confidence interval from 5.5 to 8.5 months based on the F-test.

Figure 6: Monthly perinatal mortality rates in Germany and regression line with seasonal variations.

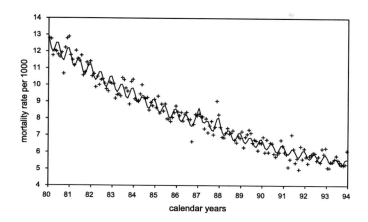

Infant mortality in Poland

For Poland, monthly data were obtained only for infant mortality (death within the first year of life) and for a relatively short period from 1985 to 1990.

Again, regressions without and with the caesium term are performed, using a time-lag of 7 months as found in the German perinatal mortality data. The corresponding sums of squares are $\chi^2 = 126.8$ (df=65) and $\chi^2 = 99.1$ (df=63). The corresponding F-test is highly significant (p=0.0004, F-test). For time-lags of both 8 and 6 months, the sums of squares are greater than for 7 months. Thus, like in the German data, 7 months is the best estimate for the time-lag. Figure 9 shows the trend of the data and the regression line; Figure 10 displays the deviations of the observed data from the expected trend (standardised residuals).

Combined regression
For a comparison of the effect size in Poland and Germany, early neonatal mortality data from Germany were used because perinatal mortality also includes stillbirths. More than 50% of infant deaths in the first year of life actually occur within the first 7 days (early neonatal deaths). To increase the precision of the parameter estimates, a combined regression with individual parameters for the long-term trend and common parameters for the seasonal components and the caesium term is performed.

The sum of squares resulting from the combined regressiom is 356.8 (df=230) without and 310.5 (df=228) the caesium term. The corresponding F-test is highly significant (p<0.0001). For time-lags of 6 as well as 8 months, the weighted sums of squares are significantly greater than for 7 months (319.0 and 337.8, respectively).

The best estimate of the power of dose is $\beta12 = 2.8 \pm 0.8$. With a linear dose response model, i.e. $\beta12=1$, the sum of squares increases significantly to 321.3 (df=229). The corresponding F-test yields p=0.0052, i.e., the dose dependency is curvilinear.

Figure 7: Deviations of monthly perinatal mortality rates in Germany from the undisturbed trend of the data (standardised residuals). The solid line is the three-month moving average, the broken lines show the 2σ-range.

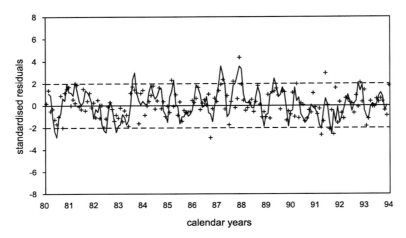

5 Belarus

The region (oblast) of Gomel was the area with highest fallout in Belarus. The strontium soil contamination in parts of Gomel oblast outside the 30-km exclusion zone exceeded 37,000 Bq/m² in 1986 (see Figure 11) whereas in Munich little strontium was determined in the Chernobyl fallout (210 Bq/m² Sr-90 compared to about 20,000 Bq/m² Cs-137, May 1986),

The trend of perinatal mortality rates, 1985-1998, in the oblasts of Gomel, Minsk-City, and in Belarus minus Gomel and Minsk-City, are displayed in Figure 12. In 1994 there is a 20% increase of perinatal mortality in all three data sets which is the consequence of a definition change for stillbirth.

Perinatal mortality data from the City of Minsk do not conform with the data in the rest of Belarus. The rates from Minsk-City are consistently higher than in the rest of Belarus until after 1995 when they suddenly drop by 50%. No such decrease occurs in the other regions of Belarus, i.e. the peculiarity in the City of Minsk, the capital of Belarus, is not likely to have biological reasons. Therefore the data for the City of Minsk are omitted when perinatal mortality rates in Gomel oblast (study area) are compared to the rates in the rest of Belarus (control area).

A trend analysis of perinatal mortality data is problematic for two reasons. First – as in most European countries - the definition of stillbirth was changed at the end of 1993. Second, possible socio-economic problems after the break-up of the Soviet Union in 1991 might have had an influence on the trend of perinatal mortality rates. Assuming that these influences acted equally in the study and the control region, a possible effect of radiation exposure should be found in the ratio of perinatal mortality rates in Gomel oblast to the rates in the control area.

Method

The ratio of perinatal mortality rate p1 in Gomel to the rate p0 in the control area can be expressed by the odds ratio (OR) which is defined by
OR = (p1/(1-p1)) / (p0/(1-p0)).

Figure 8: Sums of squares as a function of the time-lag between caesium concentration in pregnant women and perinatal mortality in Germany (profile likelihood). The broken line indicates the 95% confidence interval for the time-lag.

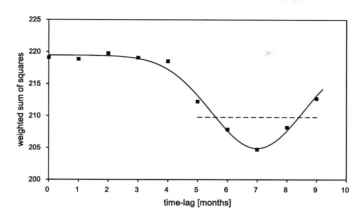

A weighted regression of the logarithms of the odds ratios is performed with the model
(5.1) $\ln(OR) = \ln(1 + \beta 0 + \beta 1 \cdot d87 + \beta 2 \cdot sr(t))$
where parameter $\beta 0$ allows for a difference in the base line perinatal mortality rates in the study and the control area, $\beta 1$ estimates a possible caesium effect in 1987 (dummy variable d87) and $\beta 2$ estimates the possible effect of strontium on the data.

The expression $sr(t)$ is the average strontium concentration in pregnant women calculated under the assumption that strontium is mainly incorporated at the end of the period of menarche, the time of maximum bone growth, at about age 14 [25].

Figure 9: Monthly infant mortality rates in Poland from 1985 to 1990 and regression line. The dotted line is the undisturbed long-term trend.

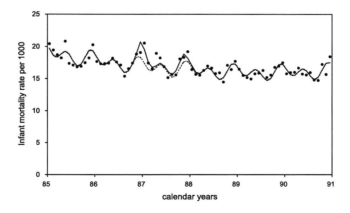

For a given year following Chernobyl, the average strontium concentration is approximated by the percentage of pregnant women who were 14 years old in 1986, i.e., who were born in 1972. This percentage follows from the maternal age distribution. In Belarus, maternal age distributions were only available in 5 year strata. The shaded area in Figure 13 is the average age distribution in Belarus for 1992-1996. To get approximated annual values, a superposition of two lognormal distributions was used to fit the step function (solid line in Figure 13).

Second, the strontium excretion from the body must be considered in the calculation of sr(t). According to the ICRP Publication 67 [26], the strontium excretion cannot be described by a simple exponential decrease, but is composed of a fast and a slow component. Thus, the strontium term sr(t) in year t after Chernobyl has the form

$$sr(t) \sim F(t-1972) \cdot (A1 \cdot \exp(-\ln(2) \cdot (t-1986)/T1) + A2 \cdot \exp(-\ln(2) \cdot (t-1986)/T2))$$

where F(t-1972) is the fraction of pregnant women born in 1972. T1=2.4 years and T2=13.7 years are effective half-lives of strontium in the female body. The constants A1, A2 and the half-lives T1, T2 were determined from a regression of tabulated values given in [27].

For the regression the data are population weighted with weights

(5.2) $\quad \sigma^2 = 1/n1 + 1/(N1-n1) + 1/n0 + 1/(N0-n0)$

where n1 and n0 are the number of perinatal deaths in the study(1) and control(0) area and N1, N0 the corresponding numbers of live births plus stillbirths.

To test the significance of the parameters, two-sided t-test are applied (H0: $\beta2=0$, $\beta3=0$).

Figure 10: Deviations of observed infant mortality rates from Poland (dots) from the undisturbed trend (standardised residuals). The solid line is the three-month moving average, the broken lines show the 2σ-range.

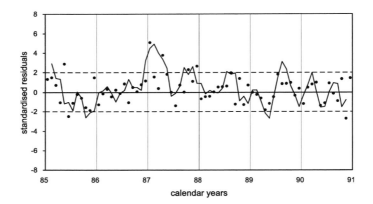

Results

To determine the age of maximum strontium uptake from the data, regressions with strontium terms for maximum strontium uptake at age 13, 14, and 15 are performed. The results for the sums of squares are 9.7, 7.3 and 9.9, respectively. Thus the strontium term with an age of 14 years for maximum strontium uptake fits the data best.

The regression yields $\beta0 = 0.022 \pm 0.027$, i.e., there is no appreciable difference in perinatal mortality rates between study and control area before Chernobyl. Also, the increase in 1987 is not significant ($\beta1=0.055 \pm 0.060$).

Not only constant $\beta2$ but also the age of maximum strontium uptake was estimated from the data therefore an F-test with two degrees of freedom is applied to determine the significance of the strontium effect. The sum of squares is 29.7 (df=12) without and 7.3 (df=10) with the strontium term. The corresponding F-test is significant (p=0.0028).

Figure 14 shows the observed (dots) and expected odds ratios (solid line) resulting from a regression with $\beta0=0$ (see equation 5.1). In the mid 1990's the odds ratios are about

1.3, i.e., perinatal mortality rates in Gomel are 30% higher than in the control area. The excess perinatal mortality rates in Gomel translate to 431 excess perinatal deaths, 1987-1998.

Figure 11: Strontium-90 soil contamination near the Chernobyl reactor site (from German journal Atomwirtschaft, March 1991). The hatched areas indicate strontium concentrations greater than 1 Ci/km² (37 kBq/m²) and 3 Ci/km² (111 kBq/m²), respectively. The circle is the 30-km evacuation zone.

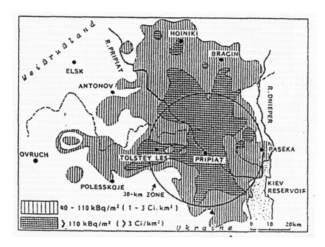

6 Spontaneous abortions in Bavaria

The most sensitive phase in embryonic development is before the first cell division, i.e., during the first hours after fertilisation. In this period an all or nothing rule holds, i.e., the fertilised egg either survives without damage or is aborted. A possible increase of spontaneous abortions will lead to a decrease of live births 9 months after exposure.

Bavaria was the German region with highest fallout. Figure 15 shows continuous measurements of the activity in air 1 meter above ground in the Munich based GSF Institute. Immediately after the Chernobyl accident the activity rose from 8 μR/h to about 110 μR/h. The number of live births plus stillbirths dropped significantly (p=0.0030) by 11.4% relative to the long-term trend in February 1987, nine months after May 1986. The decrease is more pronounced in southern Bavaria (-13,4%, p=0.0005) than in northern Bavaria (-8.7%, p=0.0370) and is limited to a single month. In March 1987, the number of births returned to the expected level. Figure 16 shows the deviations of the monthly number of live births plus stillbirths in southern Bavaria from the trend of the data, 1984-1989. The estimated number of missing births in Bavaria in February 1987 is 1154. There might have been fewer planned conceptions after the Chernobyl accident, but it does not seem plausible that the fear of an adverse pregnancy outcome was limited to May 1986.

Figure 12: Perinatal mortality rates in Gomel, Minsk-City, and Belarus minus Gomel and Minsk-City. The offset in 1994 results from a change in the definition of stillbirths.

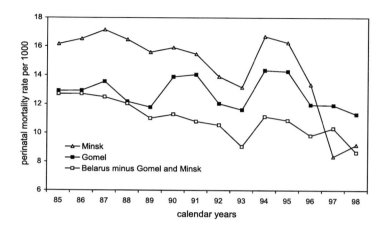

Figure 13: Maternal age distribution in Belarus, averaged for 1992-1996, and interpolation curve using two superimposed lognormal distributions.

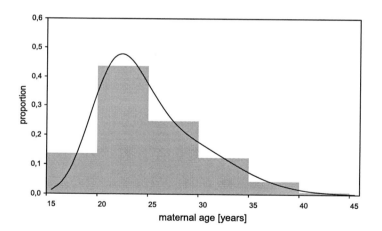

Figure 14: Odds ratios of perinatal mortality in Gomel vs. Belarus minus Gomel and Minsk-City, and regression line.

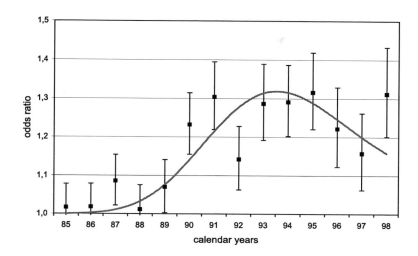

Figure 15: Gamma dose rate in the air during May 1986 in Munich, Germany. In the first days of May it reached about 110 μR/h, more than 10-times the normal level of 8 μR/h.

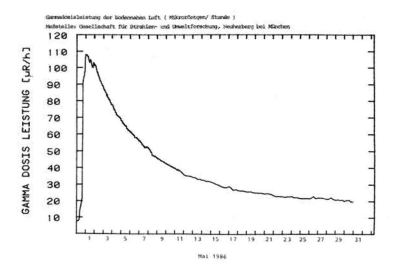

Figure 16: Deviations of the monthly number of live births plus stillbirths from the long-term trend in southern Bavaria. The broken lines indicate the 2σ-range.

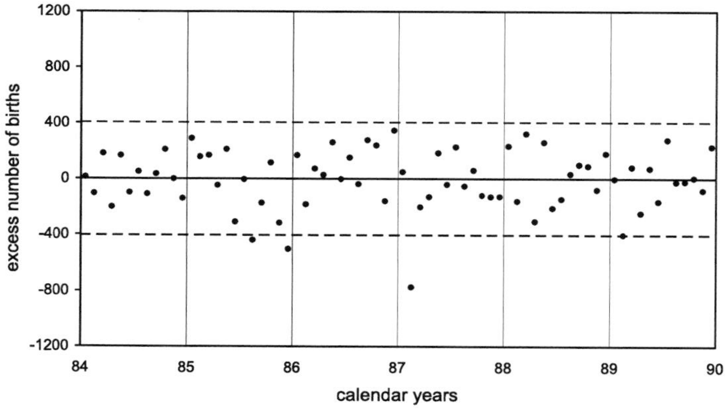

7 Discussion

A trend analysis finds significant increases of perinatal mortality in Germany and Poland in 1987, the year following the Chernobyl accident. The monthly data exhibit a significant association between perinatal mortality and the delayed caesium concentration in pregnant women. In Poland, which experienced a higher average fallout from Chernobyl than Germany, the increase of perinatal mortality in 1987, as well as the caesium effect on monthly infant mortality data, is greater than in Germany. No increase in 1987 is found in perinatal mortality data from England and Wales.

In the region of Gomel, Belarus, a significant association of perinatal mortality with the calculated strontium burden in pregnant women is found. The method used to calculate the strontium burden had already been successfully used in a trend analysis of German perinatal mortality after the atmospheric nuclear weapons tests between 1952 and 1963 [28].

In Bavaria, a significant drop in birth rate is observed in February 1987, nine months after the Chernobyl accident, which might well be the consequence of more spontaneous abortions. Similar decreases in birth rate were observed in several other European countries [4, 5, 6].

The findings provide evidence of adverse radiation effects to the foetus in the first trimester of pregnancy and challenge the widely accepted concept of a threshold dose for teratogenic effects. The foetus seems to be much more vulnerable to ionising radiation than generally believed. The form of the dose response relationship, however, is not linear.

The results for Belarus and Ukraine cannot be understood with the present dose factor for strontium. The doses from strontium in the contaminated regions as given in [29] were less than 5% of the doses from caesium but the strontium effect on perinatal mortality exceeded the caesium effect by at least a factor of 10.

The results should be interpreted with caution since they are based on aggregated data. But as long as there is no other way to study small radiation effects in human populations, the findings should not be dismissed for lack of an ultimate proof of causation.

References

Harjulehto T, Aro T, Rita H, Rytomaa T, Saxen L. The accident at Chernobyl and outcome of pregnancy in Finland. BMJ. 1989 Apr 15;298(6679):995-7.

Luning G, Scheer J, Schmidt M, Ziggel H. Early infant mortality in West Germany before and after Chernobyl. Lancet. 1989 Nov 4;2(8671):1081-3.

Lancet. 1990 Jan 20;335(8682):161-2. Comment on: Lancet. 1989 Nov 4;2(8671):1081-3. Infant mortality after Chernobyl. [No authors listed]

Ulstein M, Jensen TS, Irgens LM, Lie RT, Sivertsen E. Outcome of pregnancy in one Norwegian county 3 years prior to and 3 years subsequent to the Chernobyl accident. Acta Obstet Gynecol Scand. 1990;69(4):277-80.

Bertollini R, Di Lallo D, Mastroiacovo P, Perucci CA. Reduction of births in Italy after the Chernobyl accident. Scand J Work Environ Health. 1990 Apr;16(2):96-101.

Czeizel AE. Incidence of legal abortions and congenital abnormalities in Hungary. Biomed Pharmacother. 1991;45(6):249-54.

Bentham G. Chernobyl fallout and perinatal mortality in England and Wales. Soc Sci Med. 1991;33(4):429-34.

Little J. The Chernobyl accident, congenital anomalies and other reproductive outcomes. Paediatr Perinat Epidemiol. 1993 Apr;7(2):121-51. Review.

Kulakov VI, Sokur TN, Volobuev AI, Tzibulskaya IS, Malisheva VA, Zikin BI, Ezova LC, Belyaeva LA, Bonartzev PD, Speranskaya NV, et al. Female reproductive function in areas affected by radiation after the Chernobyl power station accident. Environ Health Perspect. 1993 Jul;101 Suppl 2:117-23.

Ericson A, Kallen B. Pregnancy outcome in Sweden after the Chernobyl accident. Environ Res. 1994 Nov;67(2):149-59.

Buzhievskaya TI, Tchaikovskaya TL, Demidova GG, Koblyanskaya GN. Selective monitoring for a Chernobyl effect on pregnancy outcome in Kiev, 1969-1989. Hum Biol. 1995 Aug;67(4):657-72.

Laziuk GI, Kirillova IA, Dubrova IuE, Novikova IV. [Incidence of developmental defects in human embryos in the territory of Byelarus after the accident at the Chernobyl nuclear power station]. Genetika. 1994 Sep;30(9):1268-73. Russian.

Petrova A, Gnedko T, Maistrova I, Zafranskaya M, Dainiak N. Morbidity in a large cohort study of children born to mothers exposed to radiation from Chernobyl. Stem Cells. 1997;15 Suppl 2:141-50.

Goldman M. The Russian radiation legacy: its integrated impact and lessons. Environ Health Perspect. 1997 Dec;105 Suppl 6:1385-91. Review.

Irl C, Schoetzau A, van Santen F, Grosche B. Birth prevalence of congenital malformations in Bavaria, Germany, after the Chernobyl accident. Eur J Epidemiol. 1995 Dec;11(6):621-5.

Grosche B, Irl C, Schoetzau A, van Santen E. Perinatal mortality in Bavaria, Germany, after the Chernobyl reactor accident. Radiat Environ Biophys. 1997 Jun;36(2):129-36.

Scherb H, Weigelt E. Comment on: Radiat Environ Biophys. 1997 Jun;36(2):129-36. A response to "Perinatal mortality in Bavaria, Germany, after the Chernobyl accident" by Grosche et al. Radiat Environ Biophys. 1998 Feb;36(4):297-9.

Korblein A, Kuchenhoff H. Perinatal mortality in Germany following the Chernobyl accident. Radiat Environ Biophys. 1997 Feb;36(1):3-7.

Scherb H, Weigelt E, Bruske-Hohlfeld I. European stillbirth proportions before and after the Chernobyl accident. Int J Epidemiol. 1999 Oct;28(5):932-40.

Scherb H, Weigelt E, Bruske-Hohlfeld I. Regression analysis of time trends in perinatal mortality in Germany 1980-1993. Environ Health Perspect. 2000 Feb;108(2):159-65.

Auvinen A, Vahteristo M, Arvela H, Suomela M, Rahola T, Hakama M, Rytomaa T. Chernobyl fallout and outcome of pregnancy in Finland. Environ Health Perspect. 2001 Feb;109(2):179-85.

Korblein A. Strontium fallout from Chernobyl and perinatal mortality in Ukraine and Belarus. Radiats Biol Radioecol. 2003 Mar-Apr;43(2):197-202.

Paretzke H. Transfer von Nukliden. Mensch und Umwelt, Magazin des GSF-Forschungszentrums für Umwelt und Gesundheit, December 1986:39-48.

Korblein A. Infant Mortality in Germany and Poland following the Chernobyl accident. Abstr 3rd Inter Conf on Health Effects of the Chernobyl Accident, 4-6 June, 2001, Kiev, Ukraine. Int J Rad Med Special Issue Vol. 3 (1-2): 63.

Tolstykh E I, Kozheurov V P, Vyushkova O V, Degteva M O. Analysis of strontium metabolism in humans on the basis of the Techa river data. Radiat Environ Biophys 1997; 36: 25-29.

INTERNATIONAL COMMISSION ON RADIOLOGICAL PROTECTION (1993). Age dependent doses to members of the public from intake of radionuclides: Part 2: Ingestion dose coefficients. ICRP Publication 67, Annals of the ICRP 23, Nos. 3-4. Pergamon Press, Oxford.

INTERNATIONAL COMMISSION ON RADIOLOGICAL PROTECTION (2003). Biological effects after prenatal irradiation (Embryo and Fetus). ICRP Publication 90, Annals of the ICRP 33, Nos. 1-2. Pergamon Press, Oxford.

Körblein A. Perinatal mortality in West Germany Following Atmospheric Nuclear Weapons Tests, Arch Envrion Health, 2006, in print.

Environmental Consequences of the Chernobyl Accident and Their Remediation: Twenty Years of Experience. Report of the UN Chernobyl Forum Expert Group "Environment" (EGE), August 2005.

243

CHAPTER 14

First Assessment of the Actual Death Toll
Attributable to the Chernobyl Disaster Based Upon Conventional Risk
Methodology

By Dr. Rosalie Bertell

IICPH, Toronto

rosaliebertell@greynun.org

Introduction

The Chernobyl disaster was in 1986, and now, 20 years after the event, there is as yet no comprehensive systematic report on the casualties. This paper presents an attempt to extend the sketchy information given in UNSCEAR 2000 (United Nations Scientific Committee of Atomic Radiation Report to the General Assembly in 2000)   to the entire population at risk and estimate the fatalities due to the radiation damage to tissue and/or its ability to initiate a fatal cancer. Many of the estimates are, of their nature, speculative, and the true estimate is most likely higher than this attempt at quantification for reasons given below. Earlier attempts have been undertaken, especially notable among them the one by Dr. John Gofman:

> *My estimate in 1986, based upon releases of various non-iodine radionuclides, was 475,000 fatal cancers plus about an equal number of additional non-fatal cases, occurring over time both inside and outside the ex-Soviet Union. Such estimates, old and new, have to be based on real-world evidence from non-Chernobyl studies --- because standard epidemiological studies (which "count" extra cancer cases) are the wrong tool for evaluating Chernobyl. No one can "see" even a half-million Chernobyl-induced cancers when they are spread among a half-billion people and occur over a century.["Don't be Fooled: 10th Anniversary of Chernobyl", Dr. John Gofman, 9 March 1996]*

This limited focus on deaths attributable to Chernobyl causes the many related tragedies of survivors to be ignored. This is especially regrettable for those many children who developed heart disease and thyroid cancers or other dysfunctions.

According to Table 1 (UNSCEAR 2000, Annex J, page 518) there were 46 radionuclides of note in the Chernobyl reactor inventory at the time of the accident. About 26 of these radionuclides were released into the air at the time of the disaster (Table 2, page 519). Of these, 17 were found in the near zone of the failed reactor (Table 6, page 521). It is important to note that the ceramic aerosolized uranium fuel particles, similar to those which have caused at least some of the devastating symptoms of the Gulf War Syndrome, were ignored in. both Table 2 and Table 6. Only radioactive cesium was used in UNSCEAR 2000 as a basis for determining the external effective radiation doses to the larger exposed population.

Information on radioactive iodine exposures is given by UNSCEAR 2000. However, since about 95% of thyroid cancers can be survived, this tragedy is not dealt with in these death estimates. Important research on radiation related heart disease in Belarus was unfortunately interrupted, and also is not included in UNSCEAR 2000.

Death due to Radiation:

The very early deaths due to radiation are those who suffered severe damage to the Central Nervous System. They die quickly, and are undoubtedly the 23 deaths that UNSCEAR 2000 attributes to the acute radiation syndrome deaths in the Moscow Hospital. Deaths due to severe exposure of the lung tissue and the red bone marrow would be expected to occur over the two years following. Given the lung and bone marrow doses in Table 14 (page 524) and the fact that 713 emergency workers (87% of 820, Table 17, page 525) had external effective doses of radiation above 0.5 Sv, I have estimated 140 deaths due to lung irradiation and 90 deaths due to bone marrow irradiation. Thus, I would estimate **deaths attributable to radiation as: 290.**

Estimated Fatal Cancers among Emergency & Recovery Workers, and the Population Evacuated and Not-evacuated from the Highly Contaminated Zones:

Emergency and Recovery Workers and accident Witnesses, were exposed to both external and internal radiation during the disaster. The estimated fatal cancer risks were based on external doses, given in Tables 16 and 17 (page 225). Cancer deaths were estimated using 10%/Person Sv, from UNSCEAR 1991 and BEIR V (U.S. National Academy Biological Effects of Ionizing Radiation) reports using the DS 86 dosimetry from Hiroshima and Nagasaki. A higher 20%/Person Sv estimate is based on the cancer risks noted in the work of Drs. John Gofman, Alice Stewart and Steve Wing, who posit risks as high as 30 to 50% per Person Sv. A risk of 20%, as used in this paper, is clearly within a reasonable probability margin of the official estimate.

Cancer deaths of Emergency and Recover Workers in 1986 are estimated to be 1,407 to 2,813. These workers undoubtedly suffered inhalation and ingestion exposure due to air contamination, and food/water contamination. Based on Table 53 (page 541), I assumed that the internal dose was about 76% of the external dose. This gives an estimate of 1,069 to 2,138. Therefore, the estimate of **cancer deaths among the emergency and rescue workers was 2,476 to 4,951. No data on the Accident Witnesses is given, contributing to the conservativeness of this estimate.**

Estimate of Cancer Deaths in the Former U.S.S.R.

Table 24 (page 528) gives the distribution of doses from external radiation of the evacuees from "areas of strict control" in Belarus. The evacuated population would be expected to have 70 to 140 cancer deaths, while those not evacuated would have 251 to 502 cancer deaths. The total number of people evacuated from the "areas of strict control" was 116,317. If those other 78.8% of evacuees had roughly the same experience as those in Belarus, their external exposure would have resulted in 260 cancer deaths. If the non-evacuated population was proportional to that in Belarus, they would have experienced 936 to 1872 cancer deaths. The UNSCEAR data failed on many scores to provide the basic information on these radiation refugees. In total, the expected **cancer deaths among those in the "areas of strict control" (the most contaminated areas, was 1,517 to 3,034.**

The cancer death estimate due to external irradiation of the former U.S.S.R. contaminated, but not controlled areas, using Table 53 (page 541) was 4,260 to 8,520. This population has received an internal radiation dose from contaminated food and water since 1986, although UNSCEAR 2000 provided no information on food and water contamination. Having no guidance from UNSCEAR 2000, I will assume that this 20 years of internal irradiation dose is about 76% of the immediate external radiation dose. Therefore the cancer deaths from internal radiation exposure would be about 3,238 to 6,475, **given a total cancer death toll of 7,498 to 3,034.**

It must be emphasized that this estimate assumes that the internal dose can be safely modeled on the basis that its effects can be predicted using the external acute exposure risk model derived from the Japanese A-Bombs. This assumption is not likely to be true for many of the isotopes involved due to anisotropy of exposure, and for this reason, the true yield from internal irradiation is likely to be some two orders of magnitude higher. The matter has been discussed by the European Committee on Radiation Risk ECRR2003 report. I

Estimated Cancer Deaths in Europe Due to the Chernobyl Disaster:

For five European countries, Croatia, Greece, Hungary, Poland and Turkey, no estimate of radiation exposure was given except for radioactive iodine dose to the thyroid. It is obvious that these countries also suffered direct external irradiation from fallout and contamination of their food and water. I have no estimates for these exposures. Eight countries, Belarus, Finland, Germany (Bavaria), Greece, Hungary, Romania, Sweden and Turkey (Black Sea Coast and Edime Province), had estimated absorbed dose given in mGy, seemingly for an epidemiological study of leukemia. Since it was a matter of leukemia, it would be normal to asses the dose to red bone marrow. No conversion of mGy to mSv was given, and the proportions of alpha (RBE 20) or beta (REB 1.7) radiation were not given. Assuming that they only calculated external radiation dose due to radioactive cesium, and based on their 1986 population, this European subgroup will be expected to have 1,517 to 3,034 cancer deaths. If we include the whole population of Europe in 1986 and assume 1 mSv effective dose per person to all not in this subgroup, from the Chernobyl fallout, we can estimate 887,819 to 1,775,638 fatal cancers. Any over estimation would be compensated for by the exclusion of consideration of contaminated food and water. **The conservative estimate of cancer fatalities in Europe attributable to Chernobyl is 889,336 to 1,778,672.**

Summary:

Using conservative methodology based on the external radiation risk factors deduced from the Japanese A-Bomb studies, I would estimate that the eventual death toll from the Chernobyl disaster would be:

> **290 due to direct radiation damage**
> **899,310 to 1,786,657 due to fatal cancers**
>
> **899,600 to 1,787,000 in total**

To repeat, this estimate is conservative for several reasons, firstly, because of the failure of the radiation investigation to document the extent of radiation contamination of food, and the absence of a comprehensive scientific examination of all deaths among the emergency and rescue workers and other populations exposed. The researchers appear to have relied on elimination of all cancers occurring in the first ten years after the accident, and a rough estimate (using a minimal risk factor reduced by a DDRF) for estimating cancer deaths. It is well known that radiation, through its mutation ability, can accelerate the development of any cancers present in

the population at the time of the disaster. Many early, uncounted cancers may fit into this category.

Again, there is also currently a scientific dispute about the acceptability of the ICRP (International Commission on Radiation Protection) methodology for assessing internal dose, especially from ceramic aerosol nuclear fuel particles, and for certain internal nuclides which bind to DNA as articulated by the ECRR (European Committee on Radiation Risk) and now accepted by the IRSN radiation protection committee in France. These particles do not spread homogeneously in the internal organs and anisotropy of dose can be very great. The UNSCEAR 2000 analysis ignored these considerations in their analysis. When this scientific effort develops an acceptable alternative to ICRP methodology, we may be able to adjust this estimate of cancer death accordingly.

Clearly, the true damage to health attributable to the Chernobyl disaster has been kept from the general public through poor and incomplete scientific investigation.

The European Committee on Radiation Risk ECRR
Greta Bengtsson

The European Committee on Radiation Risk was formed in 1997 following a resolution made at a conference in Brussels arranged by the Green Group in the European Parliament. The meeting was called specifically to discuss the details of the Directive Euratom 96/29, now known as the Basic Safety Standards Directive. This document had been passed by the Council of Ministers without significant amendment by the Parliament, and contained a statutory framework for the recycling of radioactive waste into consumer goods so long as the concentrations of itemised radionuclides were below certain levels.

The Greens, who had attempted to amend the draft with only limited success, were concerned about the lack of democratic control over such an important issue and wished for some scientific advice regarding the health effects which might follow the recycling of man-made radioactivity. The feeling at the meeting was that there was considerable disagreement over the health effects of low-level radiation and that this issue should be explored on a formal level. To this end the meeting voted to set up a new body which they named The European Committee on Radiation Risk. The remit of this group was to investigate and ultimately report on the issue in a way that considered all the available scientific evidence. In particular, the Committee's remit was to make no assumptions whatever about preceding science and to remain independent from the previous risk assessment committees such as the International Commission on Radiological Protection (ICRP), the United Nations Scientific Committee on the Effects of Atomic Radiation (UNSCEAR), the European Commission and risk agencies in any EU member State.

The ECRR's remit is:

1. To independently estimate, based on its own evaluation of all scientific sources, in as much detail as necessary and using the most appropriate scientific framework, all of the risks arising from exposure to radiation, taking a precautionary approach.
2. To develop the best scientific predictive model of detriment following exposure to radiation, presenting observations which appear to support or challenge this model, and highlighting areas of research which are needed to further complete the picture.
3. To develop an ethical analysis and philosophical framework to form the basis of its policy recommendations, related to the state of scientific knowledge, lived experience and the Precautionary Principle.
4. To present the risks and the detriment model, with the supporting analysis, in a manner to enable and assist transparent policy decisions to be made on radiation protection of the public and the wider environment.

The committee now has more than 50 experts from many countries collaborating on the issue of radiation risk and has set up a mumber of sub-committees and groups. The committee's risk model was presented in 2003 in Brussels and is published as the ECRR2003 Recommendations: the Health effects of Ionising Radiation Exposure at Low Dose for Radiation Protection Purposes (ISBN 1897761 24 4). The report, now in its second printing, has been widely circulated and translated and published in French, Russian, Spanish and Japanese. The price of the English edition is £45 with a concession price of £15 for students/ NGOs. It is available by order from any bookseller or direct by emailing an order **admin@euradcom.org**

The ECRR is increasingly being consulted by government groups, most recently by the UK government Committee on Radioactive Waste Management (CORWM) who have asked ECRR to collaborate on a waste management health impact assessment.

The risk model presented in ECRR2003 is in accord with the evidence of increases in cancer and other ill health in areas or groups exposed to low dose radiation. Recent reports of increases in cancer in Sweden (published by Martin Tondel *et al.* (2004) or the 300% increases in breast cancer reported for Belarus by the IARC researchers and, indeed, the increases in ill health shown in the present volume, are all explained or predicted by the committees model. The model also explains the increases in cancer caused by the atmospheric weapons fallout exposures of the 1960s (Busby 1994, 2002). The volume is essential reading for anyone involved in legislation in this area or who wants to obtain a realistic and rationally based set of rules to calculate the outcome of past and future exposures.

The ECRR welcomes scientists to join its group or to comment or make suggestions. Contact Greta Bengtsson at **admin@euradcom.org**

References

Busby C C (1994) Increase in cancer in Wales unexplained' *British Medical Journal* **308** 268.

Busby C C (2002) 'High risks at low doses' *Proceedings of the 4th International Conference on the Health Effects of Low Dose Radiation; Oxford Sept 24 2002* London: British Nuclear Energy Society

Pukkala E. *et al.*, (2006) Breast Cancer in Belarus and Ukraine After the Chernobyl Accident. *Int. J. Cancer*, 27th February. www3.interscience.wiley.com/doi:10.1002/ijc.21885

Tondel M, Hjalmarsson P, Hardell L, Carlsson G and Axelson O (2004) Increase in regional total cancer incidence in North Sweden due to the Chernobyl accident *J.Epid. Commun. Health* **58** 1011-1016